Applied Computing

Springer

London
Berlin
Heidelberg
New York
Barcelona
Budapest
Hong Kong
Milan
Paris
Santa Clara
Singapore
Tokyo

The Springer-Verlag Series on Applied Computing is an advanced series of innovative textbooks that span the full range of topics in applied computing technology.

Books in the series provide a grounding in theoretical concepts in computer science alongside real-world examples of how those concepts can be applied in the development of effective computer systems.

The series should be essential reading for advanced undergraduate and postgraduate students in computing and information systems.

Books in the series are contributed by international specialist researchers and educators in applied computing who will draw together the full range of issues in their specialist area into one concise authoritative textbook.

Forthcoming titles in this series include:

Deryn Graham and Tony Barrett
Knowledge Based Image Processing

Jan Noyes and Chris Baber
Designing Systems

Derek Wills and Rob Macredie
Applied Computer Graphics

Derrick Morris, Gareth Evans, Peter Green, Colin Theaker
Object Oriented Computer Systems Engineering

Linda A. Macaulay

Requirements Engineering

With 80 figures

Springer

Linda A. Macaulay, BSc (Maths), MSc (Comp.Sc), FBCS, CEng
Department of Computation, UMIST, PO Box 88, Manchester M60 1QD, UK

Series Editors

Professor Peter J. Thomas, BA, PhD, AIMgt, MIInfSc, MBCS, MIEEE, MIEE, CEng, FRSA, FVRS
Centre for Personal Information Management, University of the West of England, Coldharbour Lane, Bristol, BS16 1QY, UK

Professor Ray J. Paul, BSc, MSc, PhD
Department of Computer Science and Information Systems at St. John's, Brunel University, Kingston Lane, Uxbridge, Middlesex UB8 3PH, UK

ISBN-13: 978-3-540-76006-1   e-ISBN-13: 978-1-4471-1005-7
DOI: 10.1007/ 978-1-4471-1005-7

British Library Cataloguing in Publication Data
Macaulay, Linda
    Requirements engineering. - (Applied computing)
    1.Systems engineering 2.Software engineering
    I.Title
    005.1

Library of Congress Cataloging-in-Publication Data
Macaulay, Linda.
        Requirements engineering / Linda Macaulay
            p. cm. -- (Applied computing)
        Includes index.

        1. Systems engineering. 2. System analysis. 3. System failures (Engineering) I. Series.
TA168.M217 1996
620'.001'1--dc20   95-51545

Typesetting: Editburo, Lewes, East Sussex
34/3830-543210 Printed on acid-free paper

To Patrick, Jon, Theresa and Christine

Preface

This book has two audiences: the practising Requirements Engineer and the advanced student of software engineering or computer science. The book is unique because it introduces latest research results and, at the same time, presents highly practical and useful techniques. This book is complementary to texts on software requirements and system Requirements Engineering because of its focus on the problems caused by the fact that Requirements Engineering involves people.

Throughout this book the author has sought to introduce the reader to a number of techniques which have not previously been included within mainstream computer science literature. The techniques chosen have been shown to work in practice in both commercial and research projects. The appendices contain step-by-step guides to particular techniques; sufficient detail is provided for readers to try the techniques for themselves.

The problem faced by the Requirements Engineer is complex, it concerns meeting the needs of the customer and at the same time meeting the needs of the designer.

From a designer's point of view, Requirements Engineering is concerned with what needs to be designed rather than how it is to be designed. Requirements Engineering is also concerned with some future situation. Before design can start, there must be some knowledge of the future situation which includes the future system. Thus, Requirements Engineering is concerned with finding out about the future situation and the associated change. It is concerned with gathering information and considering possible options, and with identifying what should be designed in order to meet some perceived future need.

From a customer's point of view the situation is different. The customer lives in an ever-changing environment, in which new challenges and opportunities, new problems and constraints, occur on an almost daily basis. Current systems will never be quite adequate for the future situation that the change will bring.

Success in Requirements Engineering may be measured in terms of success of the process – for example, that the requirements process is completed within the resources allocated, or that the requirements process results in an unambiguous, complete and consistent requirements specification. Alternatively, success may be measured in terms of the ultimate success of the system – for example, in terms of customer satisfaction, meeting organisational objectives, good product reviews or increased market share.

An important aspect of this book is the emphasis on success and on lessons to be learned from failed projects. Types of system failure are examined and the causes of failure are presented. There are illustrations from a number of real projects which have failed.

In Chapter One, Requirements Engineering and the task of the Requirements Engineer are introduced. Nine different approaches to the problem of requirements are briefly described. They are marketing, psychology and sociology, object oriented analysis, structured analysis, participative design, human factors and human–computer interaction, soft systems, quality and formal computer science.

In the introduction to Chapter Two it is suggested that the objective of the Requirements Engineering process is to specify a system which will ultimately prove to be successful. Three types of system failure are introduced: process failure, interaction failure and expectation failure. Requirements Engineering is discussed in terms of these types of system failure and the cause of each type of failure is explored and discussed in some detail. Five possible causes are identified. They are: lack of a systematic Requirements Engineering process; poor communication between people; poor management of people and resources; lack of appropriate knowledge or shared understanding; and inappropriate, incomplete or inaccurate documentation.

Each cause is discussed in some detail and the need for various Requirements Engineering techniques are identified. The chapter concludes with a 'wish list' of seventy techniques and a matrix showing a mapping from the 'wish list' to the approaches discussed in Chapter One.

The next three chapters focus on particular areas within the matrix. The areas are chosen to represent particular groupings of techniques which address common causes of system failure. Each chapter deals with a different type of failure. Chapter Three focuses on expectation failure, Chapter Four on process failure and Chapter Five on interaction failure.

In Chapter Three, one cause of expectation failure is examined: the failure to realise that the goals of a system are defined within the total context of an organisation and its political and social environment and not just in relation to technology. This problem is illustrated through an analysis of the failure of a computer aided despatch system at the London Ambulance Service. The techniques described in this chapter encourage the Requirements Engineer to consider the social, political and organisational issues as part of the requirements investigation. The techniques described belong to soft systems, participative design and human factors.

In Chapter Four, one cause of process failure is examined: that different interest groups do not communicate effectively with each other, each seeking to exert power and influence over the other. The illustrative problem situation is a study of the development of a Computer Aided Learning system in which two interest groups failed to reach agreement about the requirements despite the fact that two prototypes of the system were developed. The techniques described in this chapter encourage the Requirements Engineer to consider the importance of facilitation in the Requirements Engineering process. The techniques described belong to structured analysis, quality and human–computer interaction.

In Chapter Five, one cause of interaction failure is examined: that designers do not fully understand the work of users. A number of typical problems are quoted from a variety of case studies. The techniques described in this chapter encourage the Requirements Engineer to interact with the users. They belong to marketing, participative design and human–computer interaction.

Chapter Six presents a discussion of how the Requirements Engineer might develop a 'portfolio' of requirements techniques based on seven different scenarios. For each scenario the customer–supplier relationship is explained, a typical process model is presented and techniques are suggested for the portfolio.

Appendix A contains a step-by-step guide on how to undertake a cost–benefit assessment of the organisational impact of a technical system proposal. Appendix B contains a step-by-step guide on how to conduct Stage 1 of Cooperative Requirements Capture. Stage 1 is concerned with understanding user needs and identifying the potential for change. Appendix C contains a guide on Cooperative Evaluation which is formulated to help designers conduct evaluations of their prototypes with users.

As a result of reading this text it is anticipated that the Requirements Engineer will be better equipped, both to deal with the complexities associated with meeting the needs of customers and designers, and to avoid some of the known causes of system failure.

Acknowledgements

Acknowledgements and thanks are due to all the companies who participated in my survey. To Trevor Dobbin and David Bustard of the University of Ulster for permission use their material in Chapter One. To Bruce Robinson of Salford University for the case study, to Richard Vidgen of Salford University for the rich picture and associated text and to Ken Eason of HUSAT for permission to reproduce the diagrams in Chapter Three. To Susan Gasson of Warwick University for the case study, and to Steve Viller of Lancaster University for permission to use the material on facilitation in Chapter Four.

Special thanks to Ken Eason of HUSAT and Andrew Monk of York University for permission to reproduce the step-by-step guides in Appendices A and C.

To those who commented on versions of this text, Reza Hazemi and Brian Bennett from UMIST. Particular thanks to Bob Wood of Salford University for helpful and insightful suggestions.

Special thanks to Irene Beech for many of the drawings.

Any errors and omissions are mine.

Linda A. Macaulay

Contents

1 Introduction

OBJECTIVES

- To introduce Requirements Engineering (RE).
- To explain why a requirements stage is necessary.
- To present a definition of Requirements Engineering and to introduce a general model of the RE process.
- To discuss the qualities a Requirements Engineer.
- To describe various approaches to the problem of requirements.
- To introduce the rationale for this book.

1.1 INTRODUCTION

Requirements Engineering is concerned with what needs to be designed rather than how it is to be designed. Requirements Engineering is also concerned with some future situation. In Fig. 1.1, the inputs to system design are shown as the user's present job, and the technological options and the output are shown as the future system. Drawn in this way it seems quite straightforward.

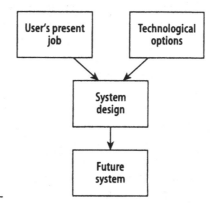

FIG. 1.1
The process of design, from the present through to future systems.

However, it is not so simple. How can the system be designed before the future situation is known? What is it that needs to be designed? In other words, what are the requirements for the design?

Before design can start, there must be some knowledge of the future situation which includes the future system. Requirements Engineering is concerned with finding out about the future situation and the associated change. It is concerned with gathering information and considering possible options, and with identifying what should be designed in order to meet some perceived future need.

In the waterfall model of system development, shown in Fig. 1.2, requirements is positioned before design. However, the arrows indicate that the requirements may not be a once-only activity, but that information which affects the requirements may be discovered at later stages in the lifecycle and that the requirements may thus need to be changed.

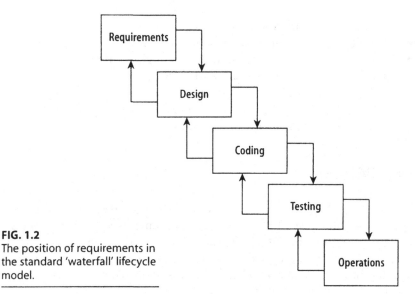

FIG. 1.2
The position of requirements in the standard 'waterfall' lifecycle model.

A slightly different view of requirements is presented in the 'V' lifecycle model, shown in Fig. 1.3. In this case, requirements is shown in two stages: the feasibility study and the product definition. The feasibility study normally involves an investigation into some customer need and the cost benefit analysis of various alternative solutions. The product definition stage is concerned with specifying the requirements for the design.

Figure 1.3 also shows an arrow going from product definition to the evaluation and acceptance of the system. This illustrates the fact that requirements do not disappear once design starts, but that the developed system must be tested against the requirements. Thus Requirements Engineering must be concerned with describing the requirements in a way that lends itself to testing.

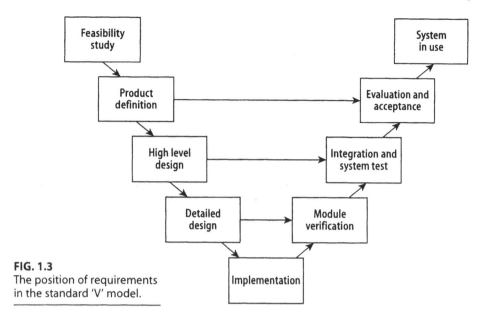

FIG. 1.3
The position of requirements
in the standard 'V' model.

1.2 WHY IS A REQUIREMENTS STAGE NECESSARY?

From a customer's point of view, the situation is different. The customer lives in an ever-changing environment, in which new challenges and opportunities, new problems and constraints, occur on an almost daily basis. Current systems will never be quite adequate for the future situation that the change will bring.

Figure 1.4 illustrates a typical system usage cycle from a customer's point of view. A customer first identifies a need for a system or product. The system is purchased and installed, users are trained how to use it, jobs may be redesigned, parts of the organisation may be restructured, and if all goes well the use of the system expands from limited usage to full usage.

Eventually, because of change in the environment, the limits of the system are reached. The customer evaluates the situation and decides that there is a need for change in the supporting system. Thus, in Fig. 1.4, the requirements stage comes under the heading of 'needs assessment'.

Of course, 'needs assessment' may not always result in the purchase and installation of a new system. It may simply be that some additional feature is needed, or an upgrade to a faster or more efficient version of the present system. In some cases it may be that the result of the needs assessment is that more effective use of personnel should be made.

Thus, from a customer's point of view, a requirements stage is necessary because it helps to understand the new needs and to identify how they can be satisfied.

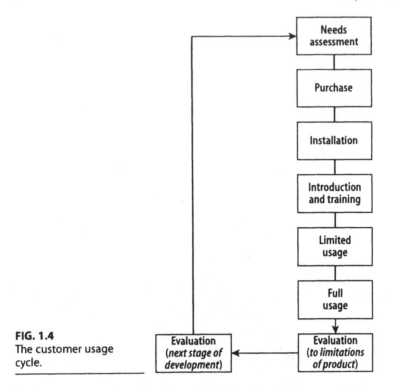

FIG. 1.4
The customer usage
cycle.

1.3 WHAT IS A REQUIREMENT?

In simple terms a requirement could be defined as 'something which a customer needs'. However, from a designer's point of view, it could also be defined as 'something which needs to be designed'. These may or may not be the same as each other.

There are a number of definitions of the term 'requirements', the most notable being IEEE (Institute of Electrical and Electronics Engineers) Standard 610 (1990):

1. A condition or capacity needed by a user to solve a problem or achieve an objective.
2. A condition or capability that must be met or possessed by a system or system component to satisfy a contract, standard, specification, or other formally imposed documents.
3. A documented representation of a condition or capability as in 1 or 2.

The IEEE standard places emphasis on software requirements, although Loucopoulos and Karakostas (1995) suggest that the definition is general enough to apply to non-software-specific situations.

In contrast to the IEEE standard the British Standards Institution (BSI) standard places the emphasis on user requirements. The BS 6719: 1986 standard guide to specifying user requirements for computer based systems does not provide a definition of requirements, but rather provides a basis for describing user needs and priorities.

Both standards recommend a contents list for a requirements document which reflects their different perspectives (see Chapter Two for further details), the requirements document itself being a statement of the requirements.

1.4 REQUIREMENTS ENGINEERING

If the requirements document is considered to be the end product of the requirements stage then Requirements Engineering can be thought of as the process by which the requirements document is populated.

Pohl (1993), in his paper on the 'three dimensions of requirements engineering', provides one of the first definitions of RE:

> Requirements Engineering can be defined as the systematic process of developing requirements through an iterative co-operative process of analysing the problem, documenting the resulting observations in a variety of representation formats, and checking the accuracy of the understanding gained.

This definition suggests that it may be rather simplistic to consider RE only in terms of a process which enables us to populate a requirements document. However, the contents lists of the various standards may well be helpful because they provide one view as to the end point of the Requirements Engineering process. (This is discussed in more detail in Chapter 2, section 2.2).

The Pohl definition is important because it highlights some of the complexities of RE. Each part of the definition leads to a number of questions:

...the systematic process of developing requirements...

- How can the process be systematic when there are many unknown factors at the beginning of the process? How can we take a step-by-step approach when we don't know how many steps are needed or when it may be unclear that the end has been reached?

...through an iterative co-operative process of analysing the problem...

- How many iterations will be needed?
- How do we know when 'enough' requirements have been gathered?

- What is meant by 'enough'?
- 'Co-operative' refers to cooperation between people. Who should be involved in the process? How are they to communicate with each other? How will they reach agreement on the process?
- Should users be active participants? Should they be involved in decision making or should they simply be sources of information?

...documenting the resulting observations in a variety of representation formats...

- What representation format should be used?
- How should the results be documented?
- What standards and which notations should be adopted and why?

...checking the accuracy of the understanding gained...

- How will we know when the process is finished?
- How accurate do the requirements need to be? What does accuracy mean in this context?
- Will everyone who is involved in the requirements process have the same understanding?

The Pohl definition was introduced in this section and a number of questions were raised concerning the RE process. These questions are revisited throughout the book. In the next section, a general model of the RE process is presented.

1.5 THE REQUIREMENTS ENGINEERING PROCESS

In his book on software requirements, Davis (1993) describes two types of activity which occur during the requirements phase of a project. .

The first is problem analysis, where 'analysts spend their time brainstorming, interviewing people who have the most knowledge about the problem on hand, and identifying all the possible constraints on the problem's solution.'

The second activity is product description, when 'it is time to take pen in hand and to make some difficult decisions and prepare a document that describes the expected external behaviour of the product...'

Davis recognises that these two activities are not necessarily sequential or mutually exclusive. While the first activity is characterised by uncertainty, an expansion of information and knowledge, the second activity is characterised by organisation of ideas, resolving conflicting views and eliminating inconsistencies and ambiguities.

Davis also recognises that the techniques employed by each activity will be quite different.

However, there is not just one single model of the process, but many.

In Chapter Six, seven different models of the RE process are suggested; though not exhaustive, they highlight the fact that different situations will require different process models.

By way of introduction, a general model of the requirements process is given in Fig. 1.5.

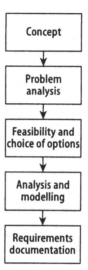

FIG. 1.5
Activities within requirements.

The product concept provides a trigger for the requirements process to begin. That trigger might be an improvement in customer service, a future need, a small incremental improvement to the existing system, or a need to use some technology which is available. Problem analysis is concerned with developing an understanding of the nature of the problem associated with the product concept. Once the problem is fully understood, possible solutions can be suggested. Feasibility and choice of options is concerned with evaluating the costs and benefits of alternative solutions. Once a given solution has been decided upon, a more detailed analysis may take place. Modelling of the application domain and of the 'solution space' may occur. Each activity should be followed by validation in order to check the accuracy of information gathered and the understanding gained. The requirements document can then be completed.

It is argued in this text that the model of the requirements process will depend to a large extent on the customer–supplier relationship. A number of different customer–supplier relationships are discussed in Chapter Six.

Thus, Requirements Engineering and the Requirements Engineering process are to some extent situation-dependent. Indeed, this is one of the reasons why it is difficult to define the task of the Requirements Engineer. Nevertheless, the next section highlights some of the engineering skills and personal qualities that are expected of a Requirements Engineer.

1.6 THE REQUIREMENTS ENGINEER

As part of the research for this book the author conducted a survey of 32 companies in order to find out more about industry's view of the Requirements Engineer (see Macaulay, 1995a, for details). The person who undertook the task of Requirements Engineering was variously referred to as a: Software Engineer; Researcher; Requirements Engineer; Requirements Analyst; Systems Analyst; Systems Engineer; Systems Engineer for Business; Requirements Modeller.

Recurring themes in the comments made by the respondents were that the Requirements Engineer should be aware of the following points:

- Requirements Engineering is not an isolated front-end activity to a software lifecycle process; rather, it is an integral part of the larger process connected to other parts through continuous feedback loops.
- The nature of requirements is such that change will necessarily occur; accordingly, requirements specifications and subsequent software lifecycles should be designed with change in mind.
- There are many sources of requirements which need to be explored and their input assimilated; there is no such thing as an all-inclusive source.
- There will be a continuous need to justify requirements and relate them to their sources; also to relate design and implementation decisions to the requirements they originate from.
- There will always be conflicts in requirements which will need to be accommodated and trade-offs that address these conflicts will need to be justified.

According to the respondents, the desirable skills that a Requirements Engineer should possess include the following:

- Interviewing skills to facilitate the acquisition of information.
- Groupwork skills, including participating in meetings and the ability to work in a collaborative way.
- Facilitation skills, such as the ability to lead a group.
- Negotiation skills to support consensus building.
- Analytical skills to support the analysis of an organisational situation prior to any proposals for solutions.
- Problem solving skills to support the search for alternative solutions.
- Presentation skills, including the ability to write coherent documents using word processors and other presentation tools.
- Modelling skills, including business, process, data and object modelling, using a variety of notations.

Finally, the fundamental and practical knowledge required for the Requirements Engineer includes:

- Knowledge of and experience in using CASE tools.
- Knowledge of and experience in using general modelling techniques, including formal languages and conceptual modelling.
- Knowledge of and experience in using particular modelling languages and the ability to choose the best one among them for a given problem.
- Knowledge of and experience with management and traceability tools.
- Knowledge of concepts, tools and techniques in human–computer interaction.
- Knowledge of techniques in product planning and marketing.

To summarise the responses, the Requirements Engineer needs the attributes of Superman or Wonderwoman!

The problem faced by the Requirements Engineer is that a wide range of knowledge and skills is needed. Consequently, a wide range of techniques may need to be employed. The next section describes potential sources of techniques.

1.7 APPROACHES TO THE PROBLEM OF REQUIREMENTS

Approaches to Requirements Engineering vary depending of the interest of the groups from which they originate. Nine groups are listed below, together with their principal interest in the problem of requirements.

1. *Marketing:* interested in the relationship between requirements and the success of a product in the market place.
2. *Psychology and Sociology:* interested in the relationship between requirements and needs of people as intelligent and social beings.
3. *Object Oriented Analysis:* interested in the relationship between requirements and the software development process, starting from a real world objects perspective.
4. *Structured Analysis:* interested in the relationship between requirements and the software development process, starting from a process and data perspective.
5. *Participative Design:* interested in requirements as part of the process of empowering users by actively involving them in the design of systems which affect their own work.
6. *Human Factors and Human–Computer Interaction (HCI):* interested in the acceptability of systems to people, the usability of systems and the relationship between requirements and evaluation of the system in use.

7. *Soft Systems:* interested in the relationship between requirements and how people work as part of an organisational system.
8. *Quality:* interested in the relationship between requirements and the quality of a product, in relation to process improvements which lead to customer satisfaction.
9. *Formal Computer Science:* interested in the relationship between requirements and software engineering's need for precision.

Each group advocates the use of specific techniques or methods within Requirements Engineering. A brief overview is given below:

1.7.1 Marketing

User support User support (sometimes known as customer support, technical support or the help desk) is a valuable source of feedback on existing products. Logs of user problems, analysis of complaints and requests for upgrades can be used to inform future releases and future products. See Keil (1995) for further details.

Surveys Surveys usually take the form of questionnaires administered to a sample of customers. Designing questionnaires and choosing appropriate sample populations is a task for professional market surveyors. The Requirements Engineer would need specific training in how to design questionnaires and how to analyse results.

User groups User groups provide an opportunity for suppliers to meet with customers to discuss the use of their products. User groups enable users from different organisations to share experiences and in this way to understand their own problems and requirements further.

Trade shows Trade shows provide an opportunity for suppliers to discuss prototypes with potential customers, and for users to experience possible future technologies.

Focus groups Focus groups could form part of a User Group meeting or of a Trade Show. The focus group involves a small group of customers coming together to have a loosely structured discussion about a particular product. The group will have a facilitator. See Morgan (1988) for further details. Focus groups are discussed in Chapter Five of this book.

Market analysis There are many market analysis techniques, such as market segmentation (see Chapter Two), market life cycles, product life cycles and competitor analysis (see QFD, Chapter Four), which can be used to identify customer requirements for computer products.

1.7.2 Psychology/Sociology

Interviewing users Interviewing is a basic tool of the Requirements Engineer. Focused interviews may be used in the early stages of the requirements investigation when the Requirements Engineer is attempting to find out about the problem domain and the concerns of users. In this case the interviewer prepares a list of topics for discussion, but not precise questions. The user is informed of the list of topics at the beginning of the interview, and each topic is explored in turn in order to elicit factual information about the user's role, the problems encountered and the functions the user performs. The interview is normally recorded on a tape recorder to facilitate later analysis. A structured interview is more appropriate when more detailed information is needed. In this case a list of specific questions is prepared. The interviewer should have a good knowledge of the problem domain and the terminology in order to adapt the questioning according to the responses. See Moser and Kalton (1971) for further details of the techniques, and Suchman and Jordan (1990) for a discussion of some of the problems of face-to-face survey interviews.

Observations In addition to asking questions, the Requirements Engineer might observe the user at work. This facilitates a deeper understanding of the environment and the day-to-day difficulties associated with the user's job. Observations need to be accompanied by some mechanism for recording the findings, either using paper and pen to keep a log of the proceedings against some predetermined proforma or through the use of video recordings. (Neale, 1989)

Video recording Video recording offers a means of studying users and their work. Sophisticated video recorders and playback facilities enable efficient and in-depth analysis of recordings – for example, through the use of time stamping and the annotation of images.

Think-aloud experiments These can be used alongside other techniques such as interviewing, user responses are recorded on audio tape. One approach is known as 'introspection', where the user is asked to describe in words how they would solve a particular problem. Another approach is to ask the user to identify and describe a 'critical incident' which occurred in the past, and to describe how it was dealt with. In contrast to this, the interviewer could think of a 'forward scenario' in which some future imaginary problem arises, and ask the user how they would deal with this. (Hart, 1985)

Card games Card games and card sorting can be used as a means of getting users to demonstrate their perception of certain types of problem. For example, the Requirements Engineer writes down the user's problems as he or she sees them, one problem per card. The Requirements Engineer then asks the user to sort the cards in order of priority. The Requirements Engineer takes the same cards and asks the user's manager to sort them in order of priority. This may help in deciding which problem to tackle first. Card sorting could also be used as a means of identifying the relationships between different concepts. In this case the cards may contain some high level concepts and some lower level concepts. The user is asked to sort the cards into piles such that there is one high level concept and the associated lower level concepts in the same pile. (Gammack and Young, 1985)

Ethnographic studies Ethnography is a method derived from anthropology and is based on observing the behaviour of groups. Professional ethnographers are employed to observe users over a long period of time and to make detailed observations about work practices. Analysis of audio-visual recordings and results from field studies 'reveal the delicate and complex web of interactional practices through which information is communicated and tasks accomplished … even apparently individual tasks like reading, writing or typing into a computer are embedded in interactions with others and are designed in relation to another's activities' (Luff *et al.*, 1993).

Their studies suggest that the traditional understanding of tasks as activities carried out by individuals is flawed and that all work activities involve some level of social interaction. Ethnography is not a requirements technique, nor is it appropriate for a Requirements Engineer to carry out an ethnographic study. However, the results of ethnographic studies can be used to inform the requirements investigation (see Hughes *et al.*, 1995).

1.7.3 Object-oriented Approaches

Object-oriented analysis is a relatively young approach to requirements analysis. According to Coad and Yourdon (1991), it is 'based upon concepts that we learned in kindergarten: objects and attributes, wholes and parts, classes and members'. One of the advantages of a specification consisting of objects is that 'it can be agreed upon more readily by all (involved) in the development process…' (Flynn, 1992)

The object-oriented approach has its origins in object-oriented programming, particularly in the language Smalltalk. The notion of object-orientation has proliferated throughout design and analysis. A recent publication of the Object Management Group (Hutt, 1994) describes twenty-one object analysis and design methods ranging from the Booch method (Booch, 1991) to Coad and Yourdon's Object Oriented Analysis (1991), Object Oriented SSADM (Berrisford, 1993) and the Z++ method (Lano and Haughton, 1992, 1993).

According to Davis (1993), the primary motivation for object orientation is that, as a system evolves, its functions tend to change but its objects remain unchanged. Thus, a system built using object-oriented techniques may be inherently more maintainable than one built using more conventional functional approaches, such as those described in the next section.

1.7.4 Structured Systems Analysis

Systems Analysis has the longest tradition of involvement in what is now called Requirements Engineering. Professional systems analysts often act as 'go betweens', assisting designers in understanding user requirements and assisting users in understanding what is possible. Their task usually involves interviewing users and collecting and analysing information from various documents. They are responsible for understanding user requirements and for describing those requirements in a form that can be understood by designers.

Systems analysts often use structured systems analysis techniques. They use dataflow diagrams to represent the processes associated with the users' tasks, and entity relationship diagrams to describe the objects that are acted upon within those processes. Data dictionaries are also used to store information about all data items defined in the dataflow diagrams; the analyst can use the data dictionary for checking the consistency and completeness of the dataflow diagrams and the entity relationship models.

These modelling techniques are useful in helping the analyst to communicate with designers because the entity relationship models can be used by the designer as a basis for the database design and the dataflow diagrams can be used as a basis for program design. Data dictionaries can be used to generate data definitions as required by languages such as COBOL, FORTRAN and PASCAL.

This approach is often referred to aa process-oriented or function-oriented. Examples of this approach can be found in:

- Structured Requirements Definition (SRD) (Orr, 1980).
- Structured Analysis and Design Technique (SADT) (Ross, 1977; Marca and McGowan, 1988).
- Structured Analysis and System Specification (SASS) (DeMarco, 1979.
- Modern Structured Analysis (McMenamin and Palmer, 1984; Ward and Mellor, 1985; Yourdon, 1989).
- Structured Systems Analysis and Design Method (SSADM) (Downs *et al.*, 1992).

Davis (1993) provides an interesting perspective on the historical development of this approach. His book also includes a review of the computer based tools which are available to support these methods. Tudor and Tudor (1995) provide an interesting comparison of structured analysis methods.

Other methods use structured approaches within a wider context of analysis (for example, CORE (Controlled Requirements Expression), Mullery, 1987). CORE undertakes the analysis by first identifying the viewpoints associated with the proposed system. A viewpoint is someone or something which is responsible for processing information. For each viewpoint a viewpoint model is developed which shows the inputs, processes and outputs using a notation similar to that of dataflow diagrams. The single viewpoint models are then connected together, outputs from one model connected to inputs to another model, to form a combined viewpoint model. This, however, is only one of a number of techniques used within CORE.

IBM's Joint Application Design (JAD) (August, 1991) also uses structured analysis techniques. JAD consists of a series of facilitated workshops and uses many different techniques in its quest for a team approach to system development. More details can be found in Chapter Four.

1.7.5 Participatory Design

Participatory Design (PD) (Floyd *et al.*, 1989; Greenbaum and Kyng, 1991) advocates a strong user involvement in system design, in which workers actively engage in designing the computer systems they will eventually use (Carmel *et al.*, 1993). PD is often referred to as the 'Scandinavian approach' because many of its proponents stem from the Scandinavian countries where the emphasis in the 1970s and 1980s was on workplace democratisation and empowering workers through participation in decision making processes.

Carmel *et al.* (1993) contrasts PD with the 'American approach': 'The software engineering approach that effectively serves as the basis for development in North America is based on fixed requirements, communication through documentation, and rules of work enforced by methods – functional foci which are de-emphasised or dismissed in the PD literature.' The PD focus is more social, with emphasis on mutual learning, joint experiences and workplace democratisation.

Greenbaum and Kyng (1991) put forward a set of design ideals which has guided their work in participatory design. The main points are listed below:

- Computer systems that are created for the workplace need to be designed with *full participation* from the users.
- When computer systems are brought into a workplace, they should *enhance* workplace skills rather than degrade or rationalise them.
- Computers are *tools*, and need to be designed to be under the control of the people using them.
- Although computers are generally acquired to increase productivity,

they also need to be looked at as a means of increasing the *quality* of the results.

- The design process is a political one and includes *conflicts* at almost every step of the way. Managers who order the system may be at odds with the workers who are going to use it. Different groups of users will need different things from the system, and the system designers often represent their own interests.
- The design process highlights the issue of how computers are used in the context of work organisation. We see this as a question of focusing on how computers are used, which we call the *use situation*, as a fundamental starting point for the design process.

Floyd *et al.* (1989) point to two main themes associated with PD principles. These are mutual reciprocal learning, and design by doing.

Mutual reciprocal learning refers to the situation in which users and designers learn from each other about work practices and technical possibilities through joint experiences. For example, designers may become 'apprentices' to users and learn by actually attempting to do the users' job. Conversely, users may learn about new technology through hands on training provided by the designer.

Design by doing refers to interactive experimentation – for example, by developing mock-ups of designs using cardboard boxes or other physical objects, games involving the future situation, or drawings of the future workplace. By employing this approach, PD practitioners have proved very innovative in engaging users in creative design.

PD practitioners (of the Scandinavian school) do not tend to be great advocates of prescriptive methods; however, they do use a number of techniques: for example:

- Future Workshops, as a means for focusing designers and users on visions of a future workplace.
- Cooperative Prototyping, in which users are actively involved in the design of prototypes as well as their evaluation.
- Design Mock-ups, for generating ideas and to get feedback from users (Ehn and Kyng, 1991).
- Future Games, an exercise in design by playing, linked to future workshops (Ehn and Kyng, 1991).

Details of these techniques can be found in Chapter Five.

Enid Mumford, who is also a PD practitioner, has adopted a more 'American approach' in that she makes available step-by-step guides on how to use the PD method ETHICS (Mumford, 1986, 1989). ETHICS places emphasis on job satisfaction, good job design and good organisational design. Users participate in the requirements process by analysing their problems at work, completing job satisfaction questionnaires and setting objectives for efficiency, effectiveness and job satisfaction.

Details of Mumford's approach can be found in Chapter Three.

1.7.6 Human Factors and HCI

In common with PD approaches, Human Factors and HCI (Human–Computer Interaction) practitioners are concerned with user centred system design. User Centred Design (UCD) places the user at the centre of the design process, and includes techniques and procedures for designing usable systems. Woodson (1981) suggests that UCD is concerned with 'design from the human-out' and that the designer should 'make the design fit the user', as opposed to 'making the user fit the design'. UCD includes techniques which cover not only the requirements stage of a project but the whole life of a product from development through to everyday usage (Rubin, 1994). Gould and Lewis (1985) advocate three key principles for UCD; these are: (i) an early focus on users and tasks; (ii) empirical measurement of product usage; and (iii) iterative design whereby a product is designed, modified, and tested repeatedly.

Eason (1992) suggests that different forms of User Centred Design are needed depending on the purpose of the product. These are shown in Fig. 1.6. When generic products are being developed for a large number of customers it is difficult to get users to participate in the design process. Eason suggests that, in this case, users are subjects who can be observed or questioned, and who can take part in user reviews or usability evaluations. When an application is being designed for a specific organisation, users can participate in the design process, probably as part of a design team. A number of alternative design team structures are discussed in Chapter Two. The third type of UCD is when users have purchased a generic product and wish to adapt it to meet their individual needs; in this case the user is in control of the changes.

UCD incorporates a whole range of user-centred techniques (Newman and Lamming, 1995), including techniques for undertaking cost benefit analyses from an organisation and user perspective, requirements capture and analysis (Bhabuta, 1989; Catterall, 1990; Macaulay, 1995c), task analysis (Johnson, 1992; Fields *et al.*, 1995), dialogue specification (Monk and Curry, 1994) and usability evaluation (Monk *et al.*, 1993; Rubin, 1994). Details of selected techniques can be found in Chapters Three, Four and Five.

1.7.7 Soft Systems

The soft systems approach recognises that all human actions take place within a wider organisational context. The work of Checkland (1981), Checkland and Scholes (1990), and Wilson (1984) demonstrates how soft systems thinking can be applied to the analysis of human and organisational problem situations.

The basic elements of the technique were establish in the early 1970s

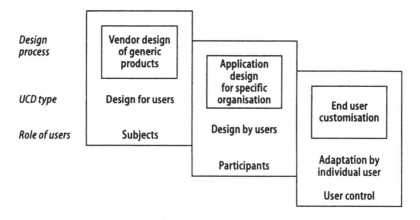

FIG. 1.6 Three forms of user-centred design (after Eason, 1992).

and refined over several years by a process of 'action research'. This involved applying the technique to real problem situations, assessing its effectiveness and making adjustments accordingly. It is not a precisely defined method 'but rather a collection of concepts that analysts may use in whatever way they find effective' (Dobbin and Bustard, 1994).

SSM is important because it has a number of features which are not explicitly addressed by more traditional analysis methods. Dobbin and Bustard (1994) present a summary of these features (reproduced with permission):

Treatment of the Problem Situation

SSM is concerned with analysing the entire problem situation, by considering the wider system environment as well as the system under investigation. SSM does not examine a problem but the situation in which there is perceived to be a problem.

Emphasis on Behaviour

SSM focuses on identifying the purpose (or purposes) of a system and the activities necessary to achieve those purposes. It explicitly avoids a consideration of system structure initially.

Emphasis on Change

SSM is a methodology which is based on the idea of bringing about change in a problem situation. The proposed system model is compared to the actual system in order to determine the necessary changes.

Multiple Perspective

The essence of SSM is its analysis of the problem situation from a number of different perspectives or viewpoints. Systems usually serve a number of different purposes and an acknowledgement of the multiple viewpoints provides SSM with a mechanism for identifying and resolving conflicts.

Goal-driven

SSM is a goal-driven approach; in other words, it focuses on a desirable system and how to reach it, rather than starting with the current situation and considering how to improve it.

Emphasis on Control and Monitoring

SSM explicitly recognises the importance of control in any system, by requiring the presence of a monitoring and control activity.

The basis of SSM is discussed in Chapter Three of this book.

1.7.8 Quality

The manufacture of mass-produced goods during the Industrial Revolution created the necessity for control of the production process, in particular in the automobile and engineering industries. Growing customer demands, intensive international competition, products which are parts or components of systems which are becoming increasingly complex, and the considerable cost savings, are all reasons put forward by Wallmuller (1994) for paying attention to quality management within software engineering.

Philip Crosby, an internationally recognised expert in the field of quality assurance, argues that not only top management but every individual is responsible for quality (Crosby, 1985). Zultner (1993) confirms that a fundamental component of total quality management is for each person or group to improve their work on a regular basis.

Zultner identifies a strong link between requirements and quality, advocating the use of QFD (Quality Function Deployment) as a means of maximising customer satisfaction from the engineering process. The focus of QFD is on 'preventing dissatisfaction by having a deeper understanding of stated requirements and implied customers' needs, and then deploying these expectations downstream in order to design value into the system.' (Zultner, 1993) Three types of requirements are identified (Kano *et al.*, 1984):

- Normal requirements: these are what we typically get by just asking customers what they want.
- Expected requirements: these are often so basic the customer may not think to mention them – until we fail to deliver them. Their presence in the system meets expectations, but does not satisfy customers. Their absence however is very dissatisfying.
- Exciting requirements: The presence of these provoke the customer reaction 'Wow. It even does this…'. Features can succeed in satisfying customers so well that they boast about their software.

Zultner describes how to use QFD and other quality control tools to improve the software development process. The basic QFD procedure is described in Chapter Four of this book.

1.7.9 Formal Computer Science

A key problem within RE is that of specifying requirements which are complete, unambiguous and consistent. Formal discrete mathematics-based specification methods have a role to play in this: for example, Lutz (1993) suggests the use of formal specification techniques alongside natural language specifications. Lutz's paper reports on the problems of software requirements errors in safety-critical embedded systems, where lack of precision and incomplete requirements have led to component failures.

Producing formal descriptions forces the Requirements Engineer to think more carefully about the nature of the system being defined and how exactly it will operate (Bustard and Lundy, 1995). Formal languages are typically difficult to learn and often are applied only to some critical subset of the specification. The reader is referred to Kent *et al.* (1993), Souquieres and Levy (1993) and Van Shouwen *et al.* (1993) for recent work on the use of formal specification languages in Requirements Engineering.

1.8 RATIONALE FOR THIS BOOK

It can be seen from the nine approaches described above that a Requirements Engineer could potentially need to learn a large number of techniques. It is not the author's intention to provide descriptions of all possible techniques. Indeed, many of the techniques mentioned above have received adequate coverage elsewhere. This book will seek to introduce the reader to a number of specific and useful techniques which have not previously been included within the Computer Science literature. The structure of this book is shown in Fig. 1.7.

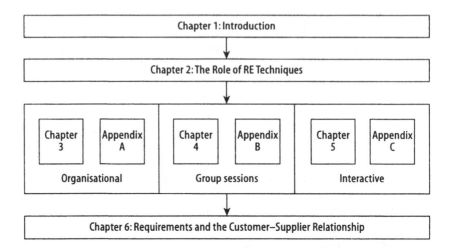

FIG. 1.7 The structure of this book.

Chapter Two presents a detailed discussion of the role of Requirements Engineering techniques and attempts to establish what techniques are needed. Requirements Engineering is discussed in terms of known types of system failure. The cause of each type of failure is explored and discussed in some detail. From this discussion a 'wish list' of seventy techniques is derived. The next three chapters focus on selected areas of the 'wish list'.

Chapters Three, Four and Five focus on specific problems within Requirements Engineering. Each chapter begins with an illustrative problem situation. There then follows a number of specific techniques which could be used to help the Requirements Engineer to deal with the problem. Also attached to each chapter is a user guide to one particular technique. It is intended that each guide contains sufficient detail for the reader to attempt to use the technique on their own project.

In Chapter Three, the key problem is that there is a failure to realise that the goals of a system are defined within the total context of an organisation and its political and social environment, not just in relation to technology. This problem is illustrated through an analysis of the failure of a computer-aided despatch system at the London Ambulance Service. The techniques described in this chapter encourage the Requirements Engineer to consider the social, political and organisational issues as part of the requirements investigation. The techniques described belong to Soft Systems, ETHICS and Eason's User Centred System Design. Appendix A contains a 'user guide' to Eason's 'Cost benefit assessment of the organisational impact of a technical system proposal'.

In Chapter Four, the key problem is that different interest groups do not communicate effectively with each other, each seeking to exert power and influence over the other. The illustrative problem situation is a study of the development of a Computer Aided Learning system in which two interest groups fail to reach agreement about the requirements, despite the fact that two prototypes of the system were developed. The techniques described in this chapter encourage the Requirements Engineer to consider the importance of facilitation in the RE process. The techniques described are JAD, QFD and CRC (Cooperative Requirements Capture). Appendix B contains a 'user guide' to Stage One of CRC.

In Chapter Five, the key problem is that designers do not fully understand the work of users. A number of typical problems are quoted from a variety of case studies. The techniques described in this chapter encourage the Requirements Engineer to interact with the users. They include future workshops, design mock-ups, cooperative prototyping and cooperative evaluation. Appendix C contains a 'user guide' to cooperative evaluation.

Chapter Six presents a discussion of how the reader might develop a 'portfolio' of requirements techniques, based upon seven different scenarios. For each scenario the customer–supplier relationship is explained, a typical process model is presented and techniques are suggested for the portfolio. This chapter also provides a conclusion to the book.

1.9 SUMMARY

In this chapter, Requirements Engineering and the task of the Requirements Engineer were introduced. Nine different approaches to the problem of requirements were briefly described. The next chapter begins with a discussion of Requirements Engineering with respect to known causes of system failure, and proceeds to derive a 'wish list' of requirements techniques.

2 The Role of Requirements Engineering Techniques

OBJECTIVES

- To discuss Requirements Engineering (RE) in relation to known causes of system failure.
- To discuss the role of RE techniques in supporting the RE process.
- To discuss the role of RE techniques in supporting the development of knowledge throughout the process.
- To discuss the role of RE techniques in supporting human communication.
- To highlight the problems associated with managing the RE process.
- To provide a 'wish list' of requirements for RE techniques which can be referred to throughout the book.
- To provide a rationale for the choice of techniques in Chapters Three, Four and Five.

2.1 INTRODUCTION

According to Pohl's definition of Requirements Engineering (RE):

> Requirements Engineering can be defined as the systematic process of developing requirements through an iterative co-operative process of analysing the problem, documenting the resulting observations in a variety of representation formats, and checking the accuracy of the understanding gained. (Pohl, 1993)

This provides a good definition of the RE process, and a starting point for considering what RE techniques[1] might be needed.

Kawalek (1993) defines a process as '...a set of identifiable, repeatable actions which are in some way ordered and contribute to the fulfilment of an objective'.

In their paper describing a 'Process Engineering Framework', Kawalek and Wastell (1994) ask what we are to expect from a 'process' and state that they are interested in processes because 'businesses need to be able to design and maintain the way they work in order that they can operate effectively and flexibly'.

In Requirements Engineering we are also interested in processes, and many recent research papers are concerned with identifying suitable processes (Bell and Oates, 1994; Dobbin and Bustard, 1994). While others are concerned with process improvement and with measuring the success of RE processes (Lubars *et al.* 1993; El Emam and Madhavji, 1995). Much of the work is concerned with identifying the 'repeatable actions which are in some way ordered' (Kawalek, 1993).

The second part of Kawalek's definition refers to process as contributing to 'the fulfilment of an objective'. What is the objective for the RE process? It could be argued that the objective is to produce a specification of requirements, this specification being some document which describes what needs to be designed. It could equally well be argued that the objective of the RE process is to specify a system which will ultimately be successful.

Bubenko (1995) recognises that most of the problems in system development have their roots not just in technical (software) issues but also in managerial, organisational, economical and social issues.

In his paper, Robinson (1994) states that as many as 50% of information systems projects may be considered to be failures. However, following his study of the failure of the systems in the London Ambulance Service, he concludes that:

> Firstly the failure or success of a project will always be defined in relation to a particular group with its own interests, roles, goals and expectations...
>
> Secondly, these interests and goals are defined in the total context of an organisation and its political and social environment and not just in relation to the technology. (Robinson, 1994)

Thus he indicates that the notion of success is not straightforward, but that it can only be defined in relation to particular groups and within the context of a particular organisational setting.

In 1987, Lyytinen and Hirschheim undertook a survey of empirical literature on information systems failures and concluded that they could be classified into four types:

1. *Correspondence failure,* when design objectives have not been met.
2. *Process failure,* which relates to the information systems development process where budget, time or other resource allocations have overrun to the point where any benefits expected from the proposed system have been negated, or where the allocated resources do not result in a workable system.

3. *Interaction failure* is the argument that a low level of use of the system can be interpreted as failure.
4. *Expectation failure* is simply that the system has failed to meet the expectations of at least one stakeholder group.

Thus, if the objective of the RE process is to specify a successful system, the Requirements Engineer needs to be aware of the possible causes of failure and must use techniques which will help avoid failure. In this chapter an attempt is made to identify what those techniques should be. The discussion which follows is based on Lyytinen and Hirschheim's classification by linking possible causes with types of failure.

Correspondence failure is not discussed, because it is argued that the purpose of the Requirements Engineering process is to set the design objectives. Thus this type of failure is not considered.

Table 2.1 shows process failure linked to three possible causes: (i) lack of a systematic RE process; (ii) poor communication between people; and (iii) poor management of people and resources. Interaction failure and expectation failure also link to three possible causes: (i) poor communication between people; (ii) lack of appropriate knowledge or shared understanding; and (iii) inappropriate, incomplete or inaccurate documentation.

TABLE 2.1 The possible causes of different types of failure.

Type of failure \ Possible cause	Lack of a systematic process	Poor communication between people	Lack of appropriate knowledge or shared understanding	Inappropriate, incomplete or inaccurate documentation	Poor management of people or resources
Process	●	●			●
Interaction		●	●	●	
Expectation		●	●	●	
Section headings	2.2 Process	2.3 Human communication	2.4 Knowledge development	2.5 Documentation	2.6 Management

Process failure is discussed under three section headings: first, the Requirements Engineering process itself; second, human communication; and third, management of the process.

Interaction failure and expectation failure are discussed under three section headings. The first is 'human communication within requirements', where it is argued that many of the problems relating to poor system use and failed expectations are due to the fact that the people involved in the RE process do not communicate sufficiently or effectively with each other. The second heading is 'knowledge development'; the problems relating to poor system use are due to a lack of knowledge about what the system should do, and, in particular, the inadequacy of the type of knowledge acquired about the users' present job, the technological options and the future situation. In addition to knowledge, people must also develop an understanding of what is needed. This knowledge and understanding will form the basis for stakeholder expectations of the system. The third heading is 'documentation of requirements'. This is included because documentation is one representation of what the proposed system should do and, as such, is a description of stakeholder expectations of the system.

Thus the remainder of this chapter is an attempt to discuss some of the problems associated with Requirements Engineering and, from these, to derive an number of requirements for RE techniques.

The problems are discussed under the following headings:

1. *The Requirements Engineering process* (a cause of process failure).

    The RE process is described as a series of activities which result in the development of intermediate workproducts.

2. *Human communication within requirements* (a cause of interaction and expectation failure).

    A key factor in specifying a successful system is understanding the needs of users and other stakeholders. This section discusses the problem of communication between users and Requirements Engineers, and the problem of identifying the needs of different groups of stakeholders.

3. *Knowledge development* (a cause of interaction and expectation failure).

    Another key factor in specifying a successful system is that all the stakeholders (including designers and Requirements Engineers) should develop appropriate knowledge and understanding. This section is concerned with the areas of knowledge that need to be developed as part of the RE process. It is concerned with abstract representations and with human understanding.

4. *Documentation of requirements* (a cause of interaction and expectation failure).

    The objective of the RE process may be a specification of requirements (for the design), but there may also be other requirements documents produced. This section discusses issues related to market requirements and to user requirements.

5. *Management of 1,2,3 and 4* (a cause of process failure).

The success of the RE process itself will also depend upon it being well managed. This section discusses some of the issues related to management.

Figure 2.1 shows the structure of this chapter.

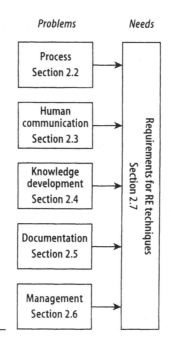

FIG. 2.1
The structure of this chapter.

The chapter concludes in Section 2.7 with a list of seventy requirements for RE techniques. These are referred to throughout the text.

2.2 THE REQUIREMENTS ENGINEERING PROCESS

As described in Chapter One, in general terms, the RE process can be thought of as a series of activities consisting of articulating the initial concept, problem analysis, feasibility and choice of options, analysis and modelling and requirements documentation. Each activity will require the use of potentially different techniques.

Often the system development process is described in terms of work products. For example, SSADM (Downs *et al.*, 1992) is described in terms of numbers of stages; each stage has a number of steps, each step has associated inputs, processes and outputs. In line with many methods the emphasis is placed on process and outputs or workproducts. This approach is attractive in situations where the development process is being audited, since workproducts can be seen and scrutinised as part of a quality assurance process.

There are a number of arguments in support of a systematic approach to Requirements Engineering. One is the ability to control the project by producing standardised maintainable outputs (Rzevski, 1980; Glasson, 1984).

A second argument is that, once a systematic procedure is in place it should be possible to measure the effectiveness of that procedure and hence to seek ways of improving it (Skousen, 1982; Wasserman, 1983).

A third argument is that the approach lends itself to the use of automated aids. Automated tools are considered important because it is assumed that they increase productivity and reduce administrative costs. (Brandt, 1983; Brodie *et al.*, 1983; Bantleman, 1985)

Thus the RE process, in general terms, consists of a series of activities. Each activity may result in a workproduct. The workproduct should be capable of being maintained and be subject to quality control. The effectiveness of the RE process should be capable of measurement, and improvements in the process should be quantifiable. The use of automated tools is desirable.

Thus RE techniques are needed which:

- support articulation of the product concept;
- support problem analysis;
- support feasibility studies and cost-benefit analyses of options;
- support analysis and modelling;
- support documentation of requirements;
- support a systematic step-by-step approach;
- provide standardised ways of describing workproducts;
- provide procedures for maintaining workproducts;
- provide ways of assessing the quality of workproducts;
- enable identification of measures and measurement of the RE process;
- support descriptions of effectiveness in RE terms;
- support analysis of opportunities for process improvement;
- provide automated support for the RE process.

This discussion has focused on the RE process as a 'set of identifiable, repeatable actions'. The next section deals with issues of who should be involved in the RE process and how they communicate with each other.

2.3 HUMAN COMMUNICATION WITHIN REQUIREMENTS

The discussion below is based on Macaulay (1993), in which a variety of issues associated with human communication are described. It is structured under the headings: (i) users are consulted; (ii) users participate; (iii) stakeholders participate; and (iv) stakeholders communicate.

2.3.1 Users Are Consulted

First, the most traditional approach is to think of the Requirements Engineer as responsible for 'eliciting' requirements from users. This is usually achieved through use of interviews, questionnaires or by observation, where the user plays a relatively passive role. In structured analysis approaches, such as SSADM (Downs *et al.*, 1992), user views are elicited at appropriate points in the method. The method relies on the expertise of the Requirements Engineer to model present activities, to elicit requirements and to develop a vision of the future to present to the project manager and other stakeholders.

Other approaches explicitly seek to identify the viewpoints which must be incorporated. For example, CORE (Mullery, 1987) requires the Requirements Engineer to identify the 'customer authority' and the 'viewpoint authority'. Each 'viewpoint authority' is the person responsible for providing the analyst with the information needed for some particular 'view' of the problem domain. That 'view' could be that of the end-user or of the user manager, or – as is often the case – the 'view' of plant, machines or controllers or some already existing computer system. The 'viewpoint' must be responsible for processing information; it must receive input from some other viewpoint and send output to a viewpoint. Thus, although a user viewpoint might be a valid viewpoint, there is a tendency for users' needs to be considered only inasmuch as they have needs as information processors.

In contrast to the data and process oriented approaches, the object oriented approach is now increasing in popularity. In particular, in Object Oriented Analysis (Coad and Yourdon, 1991), it is suggested that this approach improves analyst and problem domain expert interaction because object orientation is a natural way of thinking. The five-stage method recommended by Coad and Yourdon still prescribes a passive role for the user, with the traditional view of users as sources of information and reviewers of models developed.

The methods discussed above assume that the Requirements Engineer is responsible for understanding the problem domain and that users are a source of information and not normally active participants in deciding the requirements of the proposed system. The techniques employed within the above methods ensure that users are consulted, but do not encourage users to participate actively in the decision making process.

2.3.2 Users Participate

According to Avison and Wood-Harper (1991), in participative approaches all users are expected to contribute to and gain from any information system, and that participation should increase the likelihood of success. Participation can take many forms. For example, in ETHICS (Mumford, 1986, 1989) the users assist in the analysis of their problems at work, complete job satisfaction questionnaires, and set future objectives for efficiency, effectiveness and job satisfaction. Eason (1988, 1989), on the other hand, defines three categories of users whose needs should be taken into account: primary users, who are those likely to be frequent hands-on users of the proposed system; secondary users, who are occasional users or those who use the system through an intermediary; and tertiary users, who are those affected by the introduction of the system or who will influence its purchase but who are unlikely to be hands-on users.

In an attempt to smooth the transition from requirements to design the formation of a 'design team' is recommended. More specifically, Eason (1988) offers a number of options for the construction of the design team (whose members also have responsibility for requirements)

Evaluation criteria	Options		
	a	b	c
1. Specialist technical skills where needed	√	√	×
2. Specialist social skills where needed	×	√	√
3. Users able to contribute task knowledge	×	√	√
4. Users able to assess organisational effects	×	√	√
5. Stakeholders able to negotiate interests	×	×	√
6. All users develop feelings of ownership	×	×	√
7. Practical use of resources	√	√	×
8. Acceptable to the organisation	√	√	×

FIG. 2.2 Alternative team structures (after Eason, 1988).

where the roles of the 'technical experts' and the 'customers' are clearly identified. The technical experts contribute their skills to the creation of a system, while the customers are concerned with the world they will have to inhabit after the change caused by the new system. The customers also have a wide range of specific knowledge about the way the organisation functions and the tasks it undertakes. The technical experts will want the system to help them advance their own design skills. Eason (1988) recommends, therefore, that the structure of the design team recognises the fact that both specialists and customers have expertise to contribute and vested interests in the solutions adopted.

The three options suggested by Eason are:

a. *Technical Centred Design,* where customers commission and accept the system and are informed and consulted throughout the design process.
b. *Joint Customer-Specialist Design,* where user representatives are involved in all stages of the design process.
c. *User-Centred Design,* where the technical experts provide a technical service to the users and all users contribute to the design.

In his discussion on the alternative design team structures Eason (1988) suggests a number of criteria that could be used to evaluate the effectiveness of each structure, presented in Fig. 2.2. The first two criteria are concerned with the presence of technical skills needed and with the human and organisation specific knowledge needed if the proposed system is concerned with organisational change. Criteria three and four refer to the expert contributions that can be made by potential users, particularly the extent to which users have the opportunity to contribute specific task knowledge or to assess the organisational effects of the proposed system. Criteria five and six are concerned with the vested interests of the different stakeholders. For example, are stakeholders able to negotiate their interests and are users able to develop a feeling of ownership? The last two criteria deal with the practicality and acceptability of the design team structure as far as the commissioning organisation is concerned.

Eason's own evaluation (1988) of the alternative team structures suggests that each approach has strengths and weaknesses. The Technical Centred Design Team scores favourably for having the technical skills where needed, is a practical use of resources, and is acceptable to the commissioning organisation, but it fails on every other criteria. The Joint Customer–Specialist Design Team, on the other hand, passes on all criteria except five and six whereby stakeholders can't negotiate interests and users cannot develop a feeling of ownership. The third option, the User Centred Design Team, scores favourably on most counts. It most noticeably fails, however, on the last two criteria: it is not perceived as a practical use of resources and is generally not acceptable to the commissioning organisation.

None of the structures proposed is ideal. The Technical Centred Design Team finds favour with the commissioning organisation but largely ignores the need for participation of users and other stakeholders. The Joint Customer–Specialist Design Team is widely accepted but is likely to result in some stakeholder needs being ignored. The User-Centred Design Team, on the other hand, takes everyone's needs into account but is too inefficient in the use of resources.

User participation is widely recommended by those concerned with socio-technical design and, as can be seen from the above discussion, the ability of users to participate effectively is determined as much by the structure and remit of the design team as by the provision of suitable techniques for allowing their views to be incorporated.

2.3.3 The Stakeholders Participate

Stakeholders are defined here, using Mitroff's (1980) terms, as all those who have a stake in the change being considered, those who stand to gain from it, and those who stand to lose.

The stakeholders in any computer system fall into four distinct categories (Macaulay, 1993):

1. Those who are responsible for its design and development: for example, the project manager, software designers, communications experts, technical authors.
2. Those with a financial interest, responsible for its sale or for its purchase: for example, the business analyst, the marketing manager, the buyer.
3. Those responsible for its introduction and maintenance within an organisation: for example, training and user support staff, installation and maintenance engineers and user managers.
4. Those who have an interest in its use: for example, user managers and all classes of users, primary, secondary or tertiary.

Some of the stakeholders identified above, particularly in categories 1 and 3, have a direct responsibility for the design and development of the various system components and hence have a major interest in being involved in the requirements process. Those in category 2 have a financial responsibility for the success of the computer system and therefore may also need to be involved. The stakeholders in category 4 will be the recipients of the resulting computer system. They also have a major contribution to make in terms of specific task knowledge and the ability to assess the likely effects of the new system.

Despite Eason's observations (1988) concerning the acceptability of stakeholder participation in requirements and design, there is an increasing recognition of the need to develop a shared meaning of the system being specified and designed. For example, Konda et al. (1992)

argue that increasing design effectiveness is essentially increasing the breadth and depth of a shared meaning between the designers participating in the process of a specific design situation.

Reich *et al.* (1992) extend this notion to include not only different experts in creating this shared meaning but also a range of non-expert designers such as users, resellers and maintainers. They also argue that extracting needs from users is a dynamic, ongoing activity where the central purpose is continually to evolve the design on the basis of 'multilateral participation of all relevant actors'.

The problem of multilateral participation of a range of non-expert designers is one of potentially incompatible perspectives and conflicting objectives. Thus it is argued here that, while stakeholder participation is desirable, it will not necessarily lead to agreement or consensus as to the way forward.

2.3.4 The Stakeholders Cooperate

In this case, not only do stakeholders participate in requirements, they also cooperate with each other and are actively involved in making decisions as to the scope of the proposed new system. Similar arguments for cooperative design have been put forward by Pasch (1991) in a paper on 'Dialogical Software Design', in which he states that 'Software development is not merely a mathematical or technological challenge, but a complex social process, in which the kind of communication and cooperative, creative interaction of the participants determine the quality of the collaboratively developed product.'

He proceeds to argue that design requires experience, intuition, imagination and common sense, and that the design process is guided by insights from all participants. If this is true in software design then the case for cooperative in the requirements process can be argued even more strongly.

While it can be argued that the requirements process would be enriched by cooperation between stakeholders representative of the four categories described above, it is by no means clear that interaction between people with such a diversity of expertise and motivations (see Fig. 2.3) would result in anything but chaos. For example, Westley and Walters (1988) identify five Generic Meeting Problem Syndromes which could arise were such a requirements capture 'team' to meet to discuss requirements for a new system. These are the 'multi-headed beast', the 'feuding factions', the 'dominant species', the 'recycling meeting' and the 'sleeping meeting'. The 'multi-headed beast' syndrome, for example, can arise when there is no agreement on the agenda or when group members attempt to mix problem-solving strategies because there is no listening taking place and no integration of ideas. 'Feuding factions' occur where arguments are repeated, subgroups form within the meeting and there are hidden agendas being pursued. The 'dominant species' syndrome is often witnessed at design meetings where one member of the group attempts to dominate the rest, and other members can become withdrawn, afraid or frustrated.

Stakeholder	Motivation	Area of expertise
Software designer (1)	To produce a technically excellent system, and to use latest techniques	Latest techniques and creative design skills
Software designer (2)	To re-use existing software tools or designs	Knowledge of existing systems
Systems analyst	To produce requirements specification on time	Problem analysis
Technical author	To develop learning materials which meet user needs	Authoring skills; documentation design
User representative	To introduce change with the minimum disruption and the maximum benefit	Knowledge of the organisation, users and tasks
Training and user support staff	To support existing accounts and to generate future revenue	Knowledge of current user problems
Business/Market analyst	To be 'better' than the competition	Knowledge of business/market needs
Project manager	To complete the project successfully within given resources	Knowledge of product planning, and of previous overrun projects

FIG. 2.3 Expertise and motivation of stakeholders in a system development project.

In order for stakeholders to cooperate, therefore, it is argued that meetings need to be facilitated in some way in order that they might agree on the agenda, agree on which problem solving strategies to adopt at given points in the discussion and to ensure that all stakeholders are given the opportunity to participate.

In addition to facilitated meetings, the stakeholders need techniques which will encourage multiparty interaction and provide a focus for discussion and decision making. Examples of such techniques can be found in the QFD method (Quality Function Deployment) where a large matrix called the 'House of Quality' is drawn up by the requirements team in order to map the customer requirements on to the proposed product characteristics and features (Sullivan, 1986). The strength of the relationship between what the customer wants and what the supplier is intending to provide is entered in each cell of the matrix. Further analyses can be undertaken using the matrix as the focus for prioritising requirements and competitor analysis. Reports on the usage of QFD claim that it encourages interaction and helps build consensus and shared team understanding (Burrows, 1991). Further examples of similar techniques which encourage human communication can be found in the HUFIT toolset (Taylor, 1990), JAD (Joint Application Design) workshops (August, 1991) and Cooperative Requirements Capture workshops (Macaulay, 1994).

Thus the human communication problem concerns not only selection of personnel but also the means by which the people communicate with each other.

Thus techniques are needed which:

- provide guidance on interviewing users;
- provide guidance on the design and use of questionnaires;
- provide guidance on conducting observations of users;
- support identification of various viewpoints;
- support reconciliation of viewpoints;
- support the user in reviewing models developed;
- support users in analysing their own problems and identifying the need for change;
- support construction of appropriate requirements teams;
- support identification of stakeholders;
- support the development of a 'shared meaning' of the system being specified;
- encourage intuition, imagination and common sense among participants;
- support communication between people from a diversity of backgrounds;
- support facilitated meetings with predefined agendas and problem solving strategies;
- support the development of listening skills among participants.

Communication between people is important throughout the RE process in order to produce the documents and other workproducts required. Another important aspect of the RE process is the development of knowledge and understanding among participants. The next section discusses the areas of knowledge that need to be developed.

2.4 KNOWLEDGE DEVELOPMENT

The results of the development process are, according to Kensing and Munk-Madsen (1993), 'a system and a complete technical and organisational implementation process'.

Areas of knowledge	*Abstract knowledge*	*Concrete experience*
Users' present work	**1** Relevant structures on users' present work *Users and Developers need*	**4** Concrete experience with users' present work *Users have, Developers need*
New systems	**2** Visions and design proposals *Users and Developers need*	**5** Concrete experience with new system *Users need*
Techno-logical options	**3** Overview of technological options *Developers need*	**6** Concrete experience with technological options *Developers have, Users need*

FIG. 2.4 Six areas of knowledge (adapted from Kensing and Munk-Madsen, 1993).

They argue that the intermediate results consist not only of documents, but also of knowledge obtained by the participants, and that, regardless of the development model, be it waterfall, spiral, incremental or parallel, these intermediate results form the basis of important decisions.

Knowledge is developed by the people who are involved and different types of knowledge are needed. Within the requirements process a 'vision' of the future system needs to be acquired, knowledge of users' current practices is needed, as are projections of change using knowledge of the organisation and of external factors, and knowledge of skills and motivations of the targets users. Indeed, there are many areas of knowledge required.

Kensing and Munk-Madsen (1993) suggest that six areas of knowledge and understanding are needed before system development begins. These areas are based on the thesis that the main domains of discourse in design are:

- users' present work;
- technological options;
- new system.

Knowledge of these domains must be developed and integrated in order for the design process to be a success.

In addition, they suggest that 'two levels of knowledge' of each of these domains of discourse are required. These are:

1. Abstract knowledge, to get an overview of the domain of discourse.
2. Concrete experience, in order to understand that abstract knowledge.

Each of these areas of knowledge (see Fig. 2.4) needs to be developed as part of the Requirements Engineering process. At the beginning of the process some knowledge is already possessed. For example, the users have concrete experience of their present work (area 4) and the developers have concrete experience of technological options (area 6). Techniques are needed to facilitate sharing of this knowledge: sharing by users with developers (as in area 4); or sharing by developers with users (as in area 6). In other areas of knowledge, such as area 1 and 2, both users and developers need to acquire a common vision of the future system and agree on relevant abstract structures of the users' present work.

The type of technique used will influence the area of knowledge which can be developed and the nature of the communication between users and developers. Figure 2.5 shows a range of techniques and classifies them according to the six areas of knowledge described in Fig. 2.4.

The choice of techniques within any given development project will

Tools and techniques for knowledge development	Abstract			Concrete		
	1	2	3	4	5	6
Observations				●		
Interviewing users	●			●		
Developers doing users' work				●		
Videorecording				●		
Mock-ups				●	●	
Think-aloud experiments				●	●	
Drawing rich pictures	●			●		
Ethnographic studies	●			●		
Object-oriented analysis	●	●				
Event lists	●	●				
Entity-relationship diagrams	●	●				
Future workshops	●	●				
Conceptual modelling	●					
Dataflow diagrams	●	●				
Card games				●	●	
Formal language specifications	●					
Prototyping		●			●	●
Visits to other installations			●			●
Literature study			●			
Study of standard software			●			●

FIG. 2.5 Tools and techniques for knowledge development (adapted from Kensing and Munk-Madsen, 1993).

Present situation

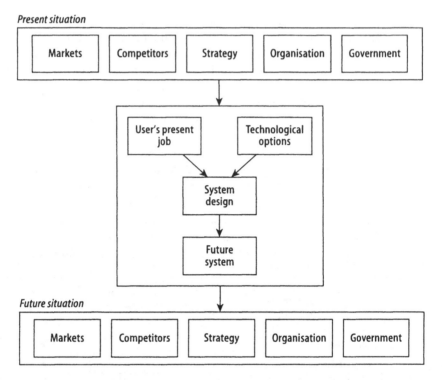

FIG. 2.6 Areas of knowledge related to Requirements Engineering.

affect the capability of the requirements team (users, developers and other stakeholders) to develop a shared understanding of the users' present work, the technological options and the future system. Ideally a range of techniques should be employed on any given project so that all six areas of knowledge are developed.

It is interesting to note that many 'traditional' requirements techniques tend to favour the development of abstract knowledge. Techniques for developing concrete experience are less well represented.

The work of Kensing and Munk-Madsen (1993) provides an excellent framework for identifying the role of requirements techniques. However, the proposed system often has strategic implications for a company and thus additional areas of knowledge may need to be developed.

Figure 2.6 illustrates the fact that, when embarking on the requirements task, it is necessary not only to develop visions of the future system but also to develop visions of the future state of markets, of competitors and competitor products, of company strategy, of planned or projected organisational change, and of course of changes in government policies and legislation.

Thus there is a need for requirements techniques which support the following developments:

- Relevant structures on the users' present work.
- Visions and design proposals.
- Overviews of technological options.
- Concrete experience with the users' present work.
- Concrete experience with the new system.
- Concrete experience with technological options.
- Knowledge of the current market and insights into future market changes.
- Knowledge of current and proposed competitor products.
- Knowledge of company strategy and likely future developments.
- Knowledge of the current organisation and the potential for change.
- Knowledge of government policy and planned changes.

How people communicate and develop knowledge are important to the success of the RE process. The next section discusses the role of the requirements document in the RE process.

2.5 THE REQUIREMENTS DOCUMENT

Most companies develop their own standard form and content of the requirements document to meet their own needs and purposes. There may be different types of requirements document. Figure 2.7, for example, shows three types, each of which contains different information.

A market requirements document would be used in situations where a supplier is wanting to develop a (generic) product which potentially can be used by a large number of customers – for example, a point-of-sale terminal for use in supermarkets and chain stores.

A typical market segmentation statement will contain the following:

1. *Industry type*: a description of the industrial groupings or types of company targeted by the proposed product.
2. *Size*: a description of the size of the market in terms of the number

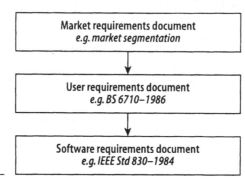

FIG. 2.7
Three different types of requirements document

Market requirements document
e.g. market segmentation

User requirements document
e.g. BS 6710–1986

Software requirements document
e.g. IEEE Std 830–1984

of buying organisations, number of sites, number of employees and number of potential product users.

3. *Geography*: a description of any geographical differences which will affect the proposed product, such as language differences, accounting systems, cultural differences, organisational structures, government regulations, and standards.

4. *Frequency*: a description of the projected frequency of purchase and frequency of use of the proposed product, including usage patterns and purchase of upgrades.

5. *End use*: a description of the likely end use of the proposed product, including a description of what users do now and how it is likely to change. A 'Day In the Life Of' a typical user or group of users is often projected.

6. *Product specification*: a description of the proposed product in terms of its objectives, its use and its impact on the customer's business.

7. *Buyer's and user's identities*: a description of the people likely to be responsible for purchase of the proposed product, including those likely to recommend its purchase within the customer organisations.

8. *Source loyalty*: a description of customer loyalty to existing work practices, to existing suppliers or other factors which may affect the acceptance of the proposed product into the organisation.

9. *Buyer's and user's personality*: information which will assist in the marketing of the proposed product, including style of promotional messages and sales channel.

In contrast to this, a user requirements document would be used by a customer who wants to describe their requirements for a system. It may be that the system will be an off-the-shelf package, a package which will need to be modified to meet the requirements, enhancements to existing systems or a bespoke system.

An example of a typical user requirements document is the BSI's (1986) *British Standard Guide to specifying user requirements for a computer-based system* (BS 6710). The standard is in two parts. The first part is a generalised summary of requirements which will introduce potential suppliers to the customer's organisation, the nature of the problem, and the principle constraints, such as cost, time and security. The second part specifies all the functions the system is required to perform, when, and in what sequence. It also includes details of acceptance criteria and performance monitoring. The BSI guide also offers advice on how to evaluate supplier responses.

Other examples of recommended contents lists for a user requirements document exist in the HCI (Human Computer Interaction) literature. For example, Macaulay (1995b) suggests:

1. *Management Summary*: includes the business case and a brief description of the proposed system.

2. *The Human Requirements*: description of the objectives of the commissioning organisation; list of the stakeholders together with their objectives; list of key workgroups and users and their objectives.

3. *The High Level Functional Requirements*: list of work roles to be supported and why; description of each work role in terms of users, objects and tasks.

4. *The Detailed Functional Requirements*: consolidated list of objects to be supported; descriptions of each object, together with details of user tasks associated with each object.

5. *The Quality Attributes*: usability, reliability, portability, performance, security, maintainability, acceptability or others, depending on the proposed system.

6. *Organisation and User Assistance Requirements*: user documentation requirements; training requirements; user support; human–computer interface requirements.

7. *The Technological Requirements and Constraints*: known hardware requirements (user or supplier); known software constraints (user or supplier).

An example of a software requirements document is the IEEE Standard 830–1984, taken from the IEEE *Guide to Software Requirements Specifications* (IEEE, 1984). A typical document's table of contents is shown in Table 2.2.

TABLE 2.2 IEEE table of contents.

1. Introduction	*n*
1.1 Purpose	**3.2** External Interface Requirements
1.2 Scope	**3.2.1** User Interfaces
1.3 Definitions, Acronyms, and	**3.2.2** Hardware Interfaces
Abbreviations	**3.2.3** Software Interfaces
1.4 References	**3.2.4** Communication Interfaces
1.5 Overview	**3.3** Performance Requirements
2. General Description	**3.4** Design Constraints
2.1 Product Perspective	**3.4.1** Standards Compliance
2.2 Product Functions	**3.4.2** Hardware Limitations
2.3 User Characteristics	...
2.4 General Constraints	**3.5** Attributes
2.5 Assumptions and Dependencies	**3.5.1** Security
3. Specific Requirements	**3.5.2** Maintainability
3.1 Functional Requirements	...
3.1.1 Functional Requirement 1	**3.6** Other Requirements
3.1.1.1 Introduction	**3.6.1** Database
3.1.1.2 Inputs	**3.6.2** Operations
3.1.1.3 Processing	**3.6.3** Site Adaptation
3.1.1.4 Outputs	...
3.1.2 Functional Requirement 2	Appendices
...	Index
3.1.*n* Functional Requirement	

A Requirements Document is a specification of what a computer system is required to do (not how it will do it). A Requirements Document can be evaluated by its effectiveness as a means of communication, by its contents as measured against a checklist, and by the quality of the statements it contains.

According to IEEE (1984), a 'good' Requirements Document should contain statements which are unambiguous, complete, verifiable, consistent, modifiable, traceable and usable during the operation and maintenance phases. These criteria are expanded below:

Unambiguous Requirements are often written in a natural language where statements can have more than one meaning. Formal requirements languages help reduce ambiguity, because the formal language processors automatically detect many lexical, syntactic and semantic errors.

Complete The Requirements Document is complete if it includes all of the significant requirements, whether relating to functionality, performance, design constraints, attributes or external interfaces and conforms to the company standard.

Verifiable An example of a non-verifiable requirement: 'The product should have a good human interface.'
An example of a verifiable requirement: 'The system will respond to a user request within 20 secs of the user pressing the *enter* key, 80% of the time.'

Consistent Three types of conflict which can occur:
- Different terms used for the same object: for example, 'a P45' and 'a tax form' might be used to describe the same form.
- Characteristics of objects conflict: for example, in one part of the requirements document, 'a red light will indicate a fault', while in another part, 'a blue light will indicate a fault'.
- logical or temporal faults: for example, 'A follows B' in one part, 'A and B occur simultaneously' in another.

Modifiable The requirements document should have a coherent and easy-to-use organisation, with a table of contents, an index and explicit cross-referencing. Requirement statements should be non-redundant where possible.

Traceable The origin of each requirement should be clear, thus facilitating 'backward traceability' to previous decisions made, and 'forward traceability' to all documents 'spawned' from the requirements document.

Usable The requirements document should be designed such that it can be referred to and if necessary modified throughout the life of the product. It should be usable even in the operation and maintenance phases.

The IEEE guide (IEEE, 1984) states that care should be taken to distinguish between the model for the application and the model for the software which is required to implement the application model. Further, the guide recommends that, whatever type of model is used, it must be rigorously defined and kept within the domain of the requirements.

The model which is to be used as the basis for the specification is crucial, as it must integrate widely-differing types of information from users. Important characteristics that such a model (and its associated language) should possess (according to Verheijen and Van Bekkum, 1982; Sowa, 1984; Flynn *et al.*, 1986) are:

1. *A high level of abstraction* The model should be at the level of the users' views of the desired system, and should capture their concepts directly. It should not indiscriminately mix this high level information with information that is relevant to lower levels of the development process (for example, details concerning data representation or physical device implementation).
2. *Human-readablility* The language in which the model is to be expressed will be used for validating the specification: that is, for presenting the specification to users for their views on its contents. Human understandability is thus the prime concern; and there is a need to avoid the well-known problems that users experience in trying to understand formal requirements specification languages.
3. *Precision* A high-level specification language, for project scope agreement, delineating the system boundaries and naming major objects, rules and processes, is required. Further detail should be left to subsequent development phases. However, the language must be precisely defined, to reduce ambiguity, and to allow for formal integration and consistency checking methods to operate on its representational form.
4. *Specification completeness* It is important that the model captures all aspects of a specification, particularly the HCI specification and, for example, system timings for real-time applications.
5. *Mappability to later phases* A Requirements Definition phase will typically be followed by detailed analysis and design phases, and the model should therefore possess a structure suitable for mapping on to the later phases.

Thus the requirements document itself may have different roles and may have differing forms and content. However, there are qualities which the documents (and the models contained within them) should possess. Thus there is a need for RE techniques which:

- support identification of requirements for generic products;
- are capable of working alongside market analysis techniques;
- support analysis of competitive products;

- support predictions of future users and future use, and estimations of future usage;
- support generic descriptions of typical users and groups of users;
- support identification and description of current workpractices;
- support identification of constraints such as cost, time and security;
- support identification and specification of acceptance criteria;
- support identification of organisational objectives, of key stakeholders and their objectives, and of key workgroups and their objectives;
- support identification of work roles to be supported and why, and descriptions of each work role, and functional requirements to support each work role;
- support identification and specification of quality attributes, such as usability, reliability, portability, performance, security, maintainability, acceptability and so on, depending on the proposed system;
- support identification and specification of requirements for user documentation, requirements for training, requirements for user support,
- support identification and description of human–computer interface requirements,
- encourage the writing of unambiguous statements,
- encourage a complete specification to be written,
- enable the development of verifiable (measurable) requirements statements,
- encourage consistency when writing requirements statements and provide support for checking consistency,
- encourage the development of modifiable requirements documents – for example, through indexing and cross referencing;
- enable traceability of requirements both backwards in the RE stage and forwards into design: that is, traceability both of the requirements and of associated documents;
- encourage the following qualities in models developed: a high level of abstraction, human readability, precision, completeness and the ability to be mapped to later phases.

The discussion so far has focused on the RE process, on human communication, on knowledge development and on the form and content of the requirements document. The next section is concerned with the problem of managing the RE process, managing communication between people including knowledge development, and managing the requirements document.

2.6 MANAGEMENT

In Chatzoglou and Macaulay (1995), Requirements Engineering is described as 'A Project Manager's Dilemma':

> Consider the following scenario: you are asked to undertake the task of managing a new project development. You have not been involved in the 'concept' stage of the project, but as a professional Project Manager it is reasonable for you to be asked to manage the requirements stage of the project.
> The remit you are given can be summarised as follows:
>
> 1. You must get the requirements 'right'. We must build the product the user wants. Quality is all important.
> 2. Here is a document which describes the overall concept of the product. You will need to work out the detail.
> 3. You must get the requirements signed off by the marketing, production, user support and strategy managers and the customer authority.
> 4. You must complete in one month. You can have one senior and two junior analysts seconded part-time to the project, and a budget of £10,000.

The dilemma is that it appears to be impossible to achieve the first three of the above given the resources stated in pararaph four. However, the Project Manager must develop a plan of how the requirements capture 'team' should proceed.

Chatzoglou and Macaulay report on research to identify those factors which affect the project plan and in particular which will enable the Project Manager to achieve the first three of the above using the minimum of resources.

They argue that, in the Requirements Engineering process, productivity is about gathering 'enough' agreed requirements[2] in order to proceed to the design stage with the minimum cost, time and effort. Productivity in the RE process can be measured in terms of the number of people involved, the time taken in the RE process (elapsed time), the cost of the RE process, the effort in man-months and the amount of requirements gathered.

The purpose of Chatzoglou and Macaulay's research was to identify those factors which affect productivity within RE and, in addition, to identify the relationship between those factors and the RE productivity measures.

Data from 107 projects was gathered from 70 different organisations within the UK. Respondents were project managers (51%), systems analysts/designers (30%) and consultants (19%); all were experienced in requirements. Respondents were asked to answer questions about the last project they were involved in. The high rate of response to questionnaires (48%) indicates that respondents feel strongly about the requirements stage of a project. Many feel that insufficient resources are given to RE. The following are representative quotations from the responses:

> I suggest you consider, very seriously, that RE is a separate exercise – *the IT strategy is determined by the RE process.*
> In general the client organisation under-appreciated the importance of the requirements specification stage.

A project manager who worked for a large company and who participated in a large project[3] stated:

> ...The project ... did not really have a properly identified RE stage. *Many problems came from this!*
> If the project had included a proper RE stage the overall goals of the project would have been substantially reduced as it would have become clear that certain 'requirements' were very ill defined, and that, in order to fulfil them, a great deal of work would have been required to properly characterise them.
> Reducing the project goals would have substantially reduced design time and would have resulted in a product brought to market more timely. It is now clear that certain of the 'requirements' were of little value.

Chatzoglou and Macaulay found that project management, project team members' attitude, and users' participation and communication with team members, were very important for the success of the RE process in most of the projects. The key factors that affect productivity in the RE process are:

1. Team members' attitude towards the systems development process, including:
   i. Team members' experience with system development and with establishing requirements.
   ii. The team's knowledge of the problem domain.
   iii. The team's commitment to and persistence with the development of the specific system.
   iv. The team's anxiety and stress for the successful completion of the RE process.

2. Users' participation and attitude towards the system development process, including:
   i. User participation in the development process and communication with team members.
   ii. User knowledge of the purpose of system development.

    iii. User motivation.

    iv. Conflicts between users.

    v. User resistance to accept the development of the specific system.

3. Project management, including:

    i. Resources available for the completion of the RE process.

    ii. Techniques and tools employed during the RE process.

    iii. Management style adopted.

4. Project characteristics, including:

    i. Project type (software, hardware, systems).

    ii. Target users (own use, external use).

    iii. Applicability (bespoke, generic).

    iv. Problem domain (well defined, moderately defined, poorly defined).

5. Other

    i. Developers (software house, industry, academics, consultants).

    ii. Information sources (customer, user, documentation, marketeer etc.).

    iii. Targets set (on time, in budget, best quality, etc.).

Management of the RE process is difficult because, as their survey also found:

1. RE is an iterative process since in only 18% of the projects just one iteration was performed. In 32% of the projects two iterations occurred, while in 50% of the projects the RE process was completed in three or more iterations.
2. The elapsed time of the RE process usually represents more than 15% of the total elapsed time; however the cost of the RE process is 5–15% of the total cost
3. The failure of 35% of the projects to capture the necessary requirements is caused by:
   - lack of time (61% of the projects);
   - poor access to information (51% of the projects);
   - insufficient manpower (22% of the projects); and
   - the cost (19% of the projects).[4]

The reasons given for changes in original plans were: lack of information; the need to validate and verify the information captured; assurance that everything has been done properly; and technology and market changes, user behaviour and inexperienced project managers.

Chatzoglou and Macaulay also found that the more time is spent in the RE stage, the less time is spent in the whole development process. Further, the higher the cost of the RE stage, the lower the cost of the whole development process. Thus the management of the RE process affects not only the RE stage itself but the whole of the project development.

Some of the issues raised above refer to the management of the RE process: the choice of the RE team, use of methodologies, resources available and management style; while others relate to knowledge development: access to users, communication with users and knowledge of the problem domain.

Other recent work is concerned with the management of the requirements document and traceability of its content into design (for example CARD (Ohnishi and Agusus), 1993; READS (Smith), 1993; RTM, 1994; DOORS, 1995).

Thus, RE techniques are needed which:

- support the management of the RE process;
- are capable of producing cost estimates of the RE process;
- are capable of modelling the RE process;
- support project planning specifically for RE projects;
- are capable of fitting with IT strategy techniques;
- help the project manager identify the skills needed to complete the RE process;
- help quantify the factors affecting productivity for a given project;
- support traceability of requirements from the Requirements Document into design and later project stages;
- support traceability of requirements from 'concept' through to the Requirements Document;
- support the management of knowledge development;
- support the management of human communication;
- support the development of a 'knowledge base' about previous projects with respect to the RE process and project success.

2.7 REQUIREMENTS FOR RE TECHNIQUES

The role of RE techniques can be summarised as being needed to support each of the areas shown in Fig. 2.8.

Requirements Engineering Techniques	
Management	The RE process
	Human Communication
	Knowledge development
	The requirements document

FIG. 2.8
The role of Requirements Engineering techniques.

The seventy requirements identified are listed below:

PROCESS TECHNIQUES WHICH:
1. Support articulation of the product concept.
2. Support problem analysis.
3. Support feasibility studies and cost-benefit analyses of options.
4. Support analysis and modelling.
5. Support documentation of requirements.
6. Support a systematic step-by-step approach.
7. Provide standardised ways of describing workproducts.
8. Provide procedures for maintaining workproducts.
9. Provide ways of assessing the quality of workproducts.
10. Enable identification of measures and measurement of the RE process.
11. Support descriptions of effectiveness in RE terms.
12. Support analysis of opportunities for process improvement.
13. Provide automated support for the RE process.

HUMAN COMMUNICATION TECHNIQUES WHICH:
14. Provide guidance on interviewing users.
15. Provide guidance on the design and use of questionnaires.
16. Provide guidance on conducting observations of users.
17. Support identification of various viewpoints.
18. Support reconciliation of viewpoints.
19. Support the user in reviewing models developed.
20. Support users in analysing their own problems and identifying the need for change.
21. Support construction of appropriate requirements teams.
22. Support identification of stakeholders.
23. Support the development of a 'shared meaning' of the system being specified.
24. Encourage intuition, imagination and common sense among participants.
25. Support communication between people from a diversity of backgrounds.
26. Support facilitated meetings with predefined agendas and problem solving strategies.
27. Support the development of listening skills among participants.

KNOWLEDGE DEVELOPMENT TECHNIQUES WHICH:
28. Support relevant structures on the users' present work.
29. Support visions and design proposals.
30. Support overviews of technological options.
31. Support concrete experience with the users' present work.
32. Support concrete experience with the new system.
33. Support concrete experience with technological options.

34. Support knowledge of the current market and insights into future market changes.
35. Support knowledge of current and proposed competitor products.
36. Support knowledge of company strategy and likely future developments.
37. Support knowledge of the current organisation and the potential for change.
38. Support knowledge of government policy and planned change.

REQUIREMENTS DOCUMENTATION TECHNIQUES WHICH:
39. Support identification of requirements for generic products.
40. Are capable of working alongside market analysis techniques.
41. Support analysis of competitive products.
42. Support predictions of future users and future use, and estimations of future usage.
43. Support generic descriptions of typical users and groups of users.
44. Support identification and description of current workpractices.
45. Support identification of constraints such as cost, time and security.
46. Support identification and specification of acceptance criteria.
47. Support identification of organisational objectives, of key stakeholders and their objectives, and of key workgroups and their objectives.
48. Support identification of work roles to be supported and why, and descriptions of each work role, and functional requirements to support each work role.
49. Support identification and specification of quality attributes: usability, reliability, portability, performance, security, maintainability, acceptability and so on, depending on the proposed system.
50. Support identification and specification of requirements for user documentation, requirements for training, requirements for user support.
51. Support identification and description of human–computer interface requirements.
52. Encourage the writing of unambiguous statements.
53. Encourage a complete specification to be written.
54. Enable the development of verifiable (measurable) requirements statements.
55. Encourage consistency when writing requirements statements and provide support for checking consistency.
56. Encourage the development of modifiable requirements documents – for example, through indexing and cross referencing.
57. Enable traceability of requirements both backwards in the RE stage and forwards into design: that is, traceability both of the requirements and of associated documents.

58. Encourage the following qualities in models developed: a high level of abstraction, human readability, precision, completeness and the ability to be mapped to later phases.

MANAGEMENT TECHNIQUES WHICH:
59. Support the management of the RE process.
60. Are capable of producing cost estimates of the RE process.
61. Are capable of modelling the RE process.
62. Support project planning specifically for RE projects.
63. Are capable of fitting with IT strategy techniques.
64. Help the project manager identify the skills needed to complete the RE process.
65. Help quantify the factors affecting productivity for a given project.
66. Support traceability of requirements from the Requirements Document into design and later project stages.
67. Support traceability of requirements from 'concept' through to the Requirements Document.
68. Support the management of knowledge development.
69. Support the management of human communication.
70. Support the development of a 'knowledge base' about previous projects with respect to the RE and project success.

This list of requirements for RE techniques is offered as an attempt to identify what techniques are needed, and why. The list is not comprehensive; some of the requirements are at different levels of detail. Nevertheless the author believes that, by developing a 'wish list' of what is needed, the Requirements Engineer will be better able to evaluate the techniques which are available.

2.8 THE CONTRIBUTION OF VARIOUS APPROACHES

In this section the author refers back to the approaches to the problem of requirements described briefly in section 1.7 and considers how each approach contributes to the requirements for RE techniques: that is, to the 'wish list'. Figure 2.9 presents a mapping from the 'wish list' to the approaches. Each '●' means that the author can identify a contribution from a particular approach to an item in the 'wish list'.

Figure 2.9 serves to illustrate the strengths of each of the nine approaches: for example, structured analysis makes a contribution to most areas but is weak on knowledge development. Participatory design approaches are strong in human communication and knowledge development but weak in almost every other area.

Figure 2.9 also serves to reinforce the view that no single approach to requirements will provide the Requirements Engineer with all the tools and techniques needed.

	wish list	market	psy & soc	OOA	struct. a	PD	HF & HCI	SSM	quality	formal
					Approaches to Requirements Engineering					
Process	1	●				●	●	●	●	
	2	●	●			●	●	●		
	3	●			●		●	●		
	4			●	●					
	5			●	●					●
	6			●	●		●			
	7			●	●					●
	8									
	9									
	10									
	11									
	12									
	13				●					
Human communication	14	●	●							
	15	●	●							
	16	●	●							
	17						●	●		
	18						●	●		
	19			●	●	●	●	●		
	20						●	●		
	21				●		●		●	
	22	●			●		●	●		
	23				●	●	●	●	●	
	24		●		●	●	●	●		
	25		●		●	●	●	●	●	
	26		●		●		●	●	●	
	27		●		●	●	●			
Knowledge development	28		●	●	●	●	●	●	●	
	29					●	●		●	
	30					●				
	31		●			●				
	32					●				
	33					●				
	34	●							●	
	35	●								
	36									
	37						●	●		
	38									
Requirements document	39	●					●		●	
	40	●							●	
	41	●					●		●	
	42	●					●		●	
	43	●	●					●		
	44				●	●	●	●		
	45				●		●	●	●	
	46						●	●		
	47						●	●		
	48						●	●		
	49				●		●	●	●	
	50						●		●	
	51			●	●	●	●		●	●
	52				●					●
	53				●					●
	54									●
	55				●					●
	56				●					
	57								●	
	58			●	●					
Management	59				●					
	60									
	61				●					
	62				●					
	63				●					
	64						●			
	65									
	66								●	
	67									
	68									
	69									
	70									

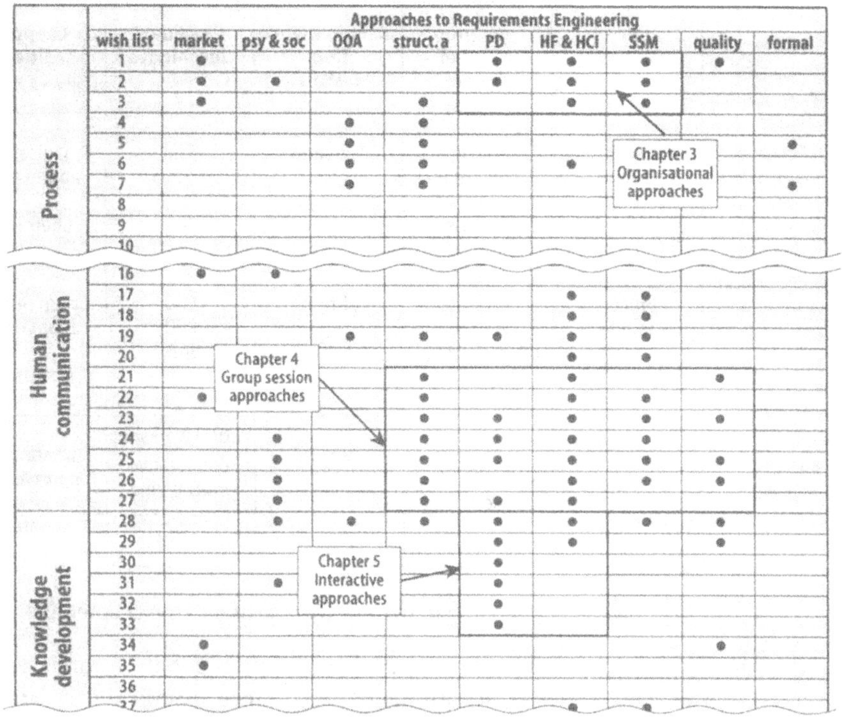

	wish list	market	psy & soc	OOA	struct. a	PD	HF & HCI	SSM	quality	formal
					Approaches to Requirements Engineering					
Process 1	•	•				•	•	•	•	
2	•	•	•			•	•	•		
3	•	•			•		•	•		
4				•	•					
5				•	•					•
6				•	•		•			
7				•	•					•
8										
9										
10										
Human communication 16		•	•							
17							•	•		
18							•	•		
19				•	•	•	•	•		
20							•	•		
21					•		•		•	
22		•			•		•	•		
23					•	•	•	•	•	
24		•			•	•	•	•		
25		•			•		•	•	•	
26		•			•		•	•	•	
27		•			•	•	•			
Knowledge development 28		•		•	•	•	•	•	•	
29						•	•		•	
30						•				
31			•			•				
32						•				
33						•				
34		•							•	
35		•								
36										
37							•	•		

FIG. 2.10 The areas addressed by the next three chapters.

It is not appropriate for a Requirements Engineering book of this type to cover all possible techniques from the above approaches. Indeed, to do so, an encyclopedia would be needed. Thus the next three chapters focus on particular areas within the matrix of Fig. 2.9. The areas were chosen because they represent particular groupings of techniques which address particular problems. Figure 2.10 shows the areas of the matrix which the next three chapters address:

In the introduction to this chapter, it was suggested that the objective of the Requirements Engineering process was to specify a system which would ultimately prove to be successful. The purpose of the next three chapters is to present examples of common causes of system failure and to suggest candidate techniques which could be used to avoid the cause of failure. Each chapter focusses on a different type of failure (as defined by Lyytinen and Hirschheim, 1987). Chapter Three focuses on expectation failure, Chapter Four on process failure, and Chapter Five on interaction failure.

◀ FIG. 2.9
The contribution of the various approaches to the requirements for RE techniques in the 'wish list'.

Type of failure	Example of cause of failure	Requirements from wish list	Candidate techniques	Chapter title
Expectation failure	social and organisational issues not addressed	1, 2 and 3	SSM ETHICS UCSD	Organisational approaches Chapter 3
Process failure	conflicts between interest groups	21 to 27	CRC QFD JAD	Group session approaches Chapter 4
Interaction failure	lack of understanding of users' work	28 to 33	Future workshop Cooperative prototyping Cooperative evaluation	Interactive approaches Chapter 5

FIG. 2.11 Mapping from the type of failure to the chapter title.

- Expectation failure can occur when insufficient attention is paid to the social and organisational context. More specifically, it can occur when there is a failure to realise that the goals of a system are defined within the total context of an organisation and its social and political environment and not just in relation to technology. (Eason, 1988; Robinson, 1994)
- Process failure can occur when different interest groups do not communicate effectively with each other, each seeking to exert power and influence over the other. (Markus and Bjorn-Anderson, 1987; Gasson, 1995)
- Interaction failure can occur when Requirements Engineers and designers do not fully understand the work of users. (Greenbaum and Kyng, 1991)

Figure 2.11 shows the relationship between the type of failure and the chapter title.

2.9 SUMMARY

In this chapter, three types of system failure were discussed, and five possible causes identified. The causes were:

- Lack of a systematic RE process.
- Poor communication between people.
- Poor management of people and resources.

- Lack of appropriate knowledge or shared understanding.
- Inappropriate, incomplete or inaccurate documentation.

Each cause was discussed in some detail, and the need for various Requirements Engineering techniques was identified. The chapter concluded with a 'wish list' of seventy techniques and a matrix showing a mapping from the 'wish list' to the approaches discussed in Chapter One. Specific areas within the matrix have been chosen as the focus for Chapters Three, Four and Five.

The next chapter, Chapter Three, concentrates on one cause of expectation failure and presents a number of requirements techniques which could be used to avoid such a cause of failure.

NOTES

1   The word 'techniques' means a method of performing or executing some specialised activity (Penguin English Dictionary) and is used throughout this book as a generic term for methods, tools, models, etc.

2   'Agreed requirements' means agreed between those people who have to 'sign off' the requirements before design commences, people such as those mentioned in paragraph 3 of the Project Manager's remit.

3   The cost of the project was around $1M; the elapsed time to completion of the project was 5 years; more than 30 people were involved; the cost and elapsed time allocated to RE process was less than 5% of the total project.

4   The sum of percentages is greater than 100% because more than one option could be chosen.

3 Specific Techniques 1: Organisational Requirements

OBJECTIVES

- To address one cause of expectation failure:
  ...that there is a failure to realise that the goals of a system are defined within the total context of an organisation and its political and social environment and not just in relation to technology... (Robinson, 1994)
- To present a case study which illustrates the problems associated with different groups of stakeholders having different system goals.
- To illustrate the key features of soft systems methodology and to discuss its role in Requirements Engineering.
- To describe ETHICS and its contribution to Requirements Engineering.
- To introduce ten principles for user-centred design.
- To present some practical techniques for cost-benefit assessment of the organisational impact of a technical system proposal.

3.1 INTRODUCTION

This chapter begins with a case study in which a technically sound system failed because the goals of the system were defined in purely technical terms. In fact, several different groups had goals for the system that were of a social, organisational or political nature. These goals were not clearly addressed as part of the specification of system requirements, resulting in the system failing to meet the expectations of at least one stakeholder group.

The main body of this chapter is dedicated to three different approaches to requirements, in which attention is drawn to the social, political and organisational context. These are the soft systems approach, participative, and user-centred design approaches. Each of these has a sound theoretical basis and the techniques they advocate have been tried and tested in real-life problem situations over many years. Each of these approaches contains techniques which can be used in a wide variety of situations, but selected techniques are introduced here because they contribute to the first three requirements from the 'wish list': that is, they are techniques which support the process of:

1. articulation of the product concept;
2. problem analysis;
3. feasibility studies and cost–benefit analysis of options.

In particular, they support process needs from a social, organisational or political perspective.

The techniques have been chosen for the following reasons:

- Soft Systems Methodology is introduced because of its emphasis on identifying the goals for a desirable system and because it provides a mechanism for identifying and incorporating different viewpoints.
- ETHICS, a participative approach, is introduced because it encourages cooptimisation of the social and technical system, and supports identification of system goals from user, management and organisational perspectives.
- Eason's approach to user-centred system design is introduced because it supports alignment of business, organisational and technical strategies and because it provides guidance on assessing the costs and benefits of alternative solutions from an organisational and user viewpoint.

3.2 AN ILLUSTRATIVE PROBLEM SITUATION

The purpose of this section is to present an analysis of the failure of a particular system and to attempt to show that even a technically excellent system could fail because of conflicting social, organisational and political goals for the system.

The analysis of failure is taken from the work of Bruce Robinson (1994), selected extracts of which are reproduced with permission.

3.2.1 The History of the Failure of the London Ambulance Service Computer-Aided Dispatch (LAS CAD) System[1]

The failure of the LAS CAD system has been one of the most spectacular and widely publicised cases of system failure in the UK in recent years. It may have caused the deaths of patients due to the late arrival of ambulances. It cost £1.5 million. It was a major embarrassment to government and LAS management. It was a matter of public concern leading to Parliamentary debate and the setting up of a Public Enquiry. We (Robinson, 1994) initially provide a brief description of the situation with the system as it existed at the point of failure. It will then be examined through the views of different interested parties.

On 26 October 1992 the full Computer Aided Dispatch system went live for the first time on an all-London basis. It aimed to create 'as far as possible a totally automated system' for dealing with emergency calls and dispatching ambulances [3012][2] by combining:

- an automatic gazetteer to locate incidents from phone numbers;
- location of ambulances through their transmission of radio signals;
- display of ambulances and incidents on controllers' displays;
- system proposal of the most suitable available resources to answer a particular call;
- radio communication of allocation to mobile data terminals in the ambulances.

On the morning of 26 October, workers in the control room were faced with a number of changes to the way in which they did their work:

- the introduction of the system including the sole use of system proposed allocation of ambulances;
- the total replacement of previous paper records;
- the absence of either paper or systems backup;
- reconfiguration and furnishing of the control room;
- removal of the previous division of London into three divisions, each with its own dedicated controllers;
- replacement of controllers by control room assistants. [4015]

Neither control staff nor ambulance crews had been adequately trained on the system and staff were working in an unfamiliar setting. [4001] On the technical side, the system was running without tested backup and 'the software was not complete, not properly tuned and not fully tested ... There were outstanding problems with data transmission to and from the mobile data terminals.' [4001] Simultaneous changes to job organisation and the work environment meant that both contributed to the failure.

Live operation of the system brought problems to light almost immediately. They were caused by the interaction of a number of operational causes which became self-perpetuating and mutually reinforcing once the system began to build up a backlog of calls. The end results were: many or no vehicles were sent to incidents; lost calls or long delays to calls; resources incorrectly shown as unavailable; and an increased number of calls to ask why ambulances hadn't arrived. As a result a number of patients waited many hours for an ambulance, including, in one case, 17 hours. Though no coroner found that death resulted from delays in arrival of ambulances, the press documented a number of cases where death had occurred following delays. [6090-2; *Daily Mirror*, 28/10/92; 29/10/92; 30/10/92]

At this point the Chief Executive of the LAS resigned and a debate took place in Parliament, during which the Health Secretary announced that LAS had set up an external inquiry into events. Meanwhile LAS had reverted to partial use of the system that was in operation before 26 October. [4034]

This worked reasonably until 4 November, when the system locked and fallback to a second system failed. The cause of the second crisis was traced to a programmer error which had been dormant for over two weeks. [4033–4041] The CAD system was then totally abandoned and LAS reverted to manual methods.

3.2.2 Four Participant Views of the LAS Failure[1]

The failure of the system was generally accepted, although conflicting explanations were given of why it had occurred. The range of people who saw themselves as having a stake in the CAD system was large. We [Robinson, 1994] looked at the views of four groups with major interests in the development of the CAD system: the top management of LAS; the union NUPE, which represented the majority of the workforce; the Technical Manager of LAS; and the Government.

Top Management

The CAD system was seen as a central part of their strategy to change the LAS culture and improve performance. The CAD system was also seen as a way of replacing the controllers (who tended to be older, traditionally minded and experienced ambulance officers on higher grades) by less skilled control room assistants. Alongside this there was an attempt to change the culture from that of a public service to something more akin to a business. For example, senior staff were taken out of uniform and put into suits. [6017] As a result of the dominance of LAS by the top management and the central role of the CAD system, when it failed it was seen as the failure of the entire strategy, which was then abandoned.

The Union (NUPE)

Management attempts to by-pass the union were the reason that no structures were set up to involve staff in systems development (NUPE, 1992). As a result, union warnings of problems were ignored and the workforce was not given any channel to participate or give feedback on their experience with the system.

The Technical Manager of LAS

The Technical Manager of LAS was quoted [in the Independent on Sunday, 8/11/92] as stating 'The computer system did not fail on that

Monday – it was working exactly as it was designed to do.' The Inquiry Report echoed this comment to some extent, but added 'much of the design had fatal flaws that would, and did, cumulatively lead to all the symptoms of system failure.' The Inquiry emphasised that the system needed to contain perfect information, which is dependent on every human in the system acting exactly as required. [4008] This means that the system designer needed to pay attention to the way people actually work.

For the Technical Manager his task ends with a system which meets its specification, is well designed and runs quickly. The Manager also stated [*Independent on Sunday*, 8/11/92]:

> What can also be expected to emerge is that the people who are using the computer system ... are 'doing things wrong'. That is, they are not doing what the computer has been designed to expect... The Inquiry may find that this is due to the computer specialists' failure to appreciate the way ambulance staff work. But this is so often erratic and ill-organised, the computer staff complain, that it is no surprise the process cannot be moulded to fit the disciplines of the automatic system.

The Government

The Government had a large stake in the success of the LAS CAD system. Ministers had used it both as an example of the amount of money being invested in LAS and to deflect criticisms of the service by insisting that the CAD system would resolve existing problems when implemented.

3.2.3 Social Interests and the Explanation of Systems Failure[1]

These four views of the LAS failure represent not merely four different interpretations of why the CAD system failed, they are also rooted in the location of their protagonists in relation to the organisation, the system and broader political issues. The protagonists differ in their goals for the organisation, goals for the CAD system and more globally in the way they saw technology. Figure 3.1 outlines how these views may be categorised.

The goals are clearly contradictory and reveal conflicts between the parties, which expressed themselves not merely generally in relation to the organisation but also in relation to the CAD system and its development. For example, the Technical Manager had conflicts both with the LAS management, who had ignored his recommendations on the CAD system, and with the staff, whose stream of suggestions for improving the system he saw as creating dangerously informal channels outside the project management structure. NUPE and LAS management held deeply antagonistic views of the CAD system's function and the development process. The centrality of the CAD system to the organisation meant that

Group	Organisational goals	Attitude to technology	Interest in the system
LAS management	Trust status Culture change Lower costs By-pass unions	Instrumental	Means of reaching goals within LAS
NUPE	More resources Better working conditions Accountability Union position maintained	Dependent on context	Means of improving working conditions and doing the job better
Systems manager	Produce and manage good systems	Functionalist/ engineering	Provision of good service and systems to LAS
Government	Good example of NHS policy Value for money Less union power	Political	Problem solver/ flak-catching device

FIG. 3.1 Four different organisational goals for the LAS system (Robinson, 1994).

it also took on a symbolic significance, which reflected the power battles going on.

3.2.4 Conclusions from the LAS CAD System[1]

A number of conclusions can be drawn from the analysis presented, which have direct relevance to the activities of both practitioners and analysts of systems development and implementation.

First, the failure or success of a project will always be defined in relation to a particular social group with its own interests, roles, goals and expectations. While agreement may occur between these groups, it should be seen as something to be explained by the specific situation they find themselves in, rather than being taken for granted.

Second, these interests and goals are defined in the total context of an organisation and its political and social environment and not just in relation to the technology. Success or failure is not just a matter of 'getting the technology right' or even of ensuring that it 'fits' an existing or planned organisational structure. Factors such as the Government's health policy may be totally outside the control of those involved in the system, but yet may play a decisive role in how people view the system.

This suggests that technologists cannot just accept or assume a given boundary between their own technical sphere and the social environment if their systems are to avoid being seen as failures (Van Lieshout and Massink, 1993). Such boundaries are in any case not fixed and subject to definition (Low and Woolgar, 1993). However, once this is accepted, the technologists then become explicitly and directly involved in all the social conflicts. The technologists' role then changes radically and the methods used must reflect this. It also implies that there may be situations where the technologists may not be able to influence whether a system is seen as a failure or not.

3.2.5 Conclusions from the Illustrative Problem Situation

A number of points can be drawn from this case study in terms of the themes of this book. First, the success of a project cannot be seen solely in terms of the success of the technology. If expectation failure is to be avoided, the goals of all the interest groups (stakeholders) must be understood from the outset. Second, the role of the Requirements Engineer in this process is unclear. However, what is clear is that the RE cannot assume a given boundary between the technical system and the organisational/social system, but needs to be involved in the articulation of that boundary. Third, the role of Requirements Engineering techniques in supporting articulation of the product concept, problem analysis and cost–benefit analysis from a stakeholder and organisation point of view is critical to this process.

The next three sections present details of specific techniques which are designed to help projects avoid expectation failure. The first technique is the Soft Systems Methodology.

3.3 SOFT SYSTEMS METHODOLOGY

The soft systems methodology (SSM) was introduced briefly in Chapter One, Section 1.7.7. The aspects of SSM which are key to the problems addressed in this chapter are:

- SSM is a goal-driven approach. In other words, it focuses on a desirable system and how to reach it, rather than starting with the current situation and considering how to improve it.
- SSM supports the analysis of the problem situation from a number of different perspectives or viewpoints.

Checkland and Scholes (1990) describe a number of case studies of the application of SSM techniques, including the use of SSM in workshop situations in which the various stakeholders are represented. In fact SSM is used in many different ways, for example McMaster *et al.* (1995) use SSM as a means of identifying the stakeholders associated with a university car park. Specific applications include information systems supporting port operations in Australia (Watson and Smith, 1988); procuring warship systems for the UK Ministry of Defence (Strain, 1990); improving tree crop agrotechnology processes in Hawaii (Mills-Packo *et al.*, 1991) and information systems for primary health care in the Aegean Islands (Darzentas and Spyrou, 1993).

The basic methodology is summarised in Fig. 3.2. It has seven distinct stages (Wilson, 1984):

1. Finding out about the problem situation.
2. Expressing the problem situation (rich picture of the real world).
3. Selection: i.e., selecting how to view the situation to produce insights and producing root definitions.
4. Building conceptual models of what the system must do for each root definition.
5. Comparison of the conceptual model with the real world.
6. Identifying feasible and desirable changes.

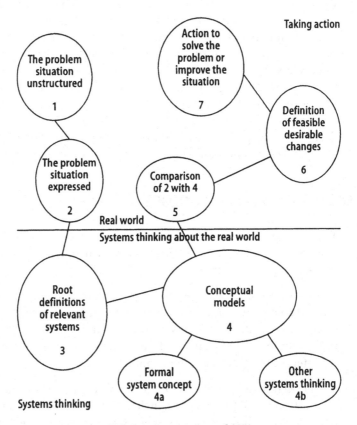

FIG. 3.2 An overview of SSM.

7. Recommendations for taking action to improve the problem situation.

It is possible to start the methodology at any stage and iteration and backtracking are recommended. It is also recommended that the stages above the line (see Fig. 3.2) are expressed in a language that is readily understood by the people involved with the problem situation. The stages below the line require a specialised systems language.

Stage 1: The problem situation: unstructured

The purpose of Stage 1 is to describe the problem situation. Information is gathered about who is involved, what their perceptions of the situation are, what the organisation structures are, and what processes are going on.

An example of a (fictitious) problem situation is given in Vidgen (1994):

> The scenario is a vehicle rental company (VCR plc). VCR rents cars and light vans to private and business users. They have noticed that there has been a significant rise in the level of business rentals – market research predicts that business rentals will be the fastest growing market sector over the next five years. VCR believe that growth in business rentals is fuelled by the following factors:
>
> - Due to the recession, organisations no longer need to offer company cars to attract and retain employees.
> - Inland Revenue taxation of company cars as a benefit in kind is thought to become increasingly punitive (VCR attribute this to pressure from environmentalists and the Government's need to raise revenue).
> - A desire to come into line with the company policies of other EC countries (company cars are virtually unheard of outside the UK)
>
> VCR is considering whether it should establish a separate corporate services operation to target medium to large organisations. VCR's strategy is to become a sole supplier of vehicle rentals to its corporate customers.

Stage 2: The problem situation: expressed

The important features of the problem situation are expressed in a way which helps relevant systems to be chosen in Stage 3. Pictorial formats are recommended – the phrase 'Building a rich picture' is often used to describe Stages 1 and 2. Greater detail may be added later to the rich picture to support Stage 5. The rich picture should show the main structures – e.g. power structure, power hierarchy, reporting structure – and the

FIG. 3.3 Rich picture of a vehicle rental company.

pattern of formal and informal communications. It should also show elements of process, thus forming a view of how structure and process relate to each other in the situation being investigated.

Vidgen provides an excellent example of a rich picture for the vehicle rental company, shown in Fig. 3.3.

Rich pictures are helpful in gaining an understanding of a situation, and they provide the Requirements Engineer or the requirements team with a basis for developing a common understanding of the situation. However, as Vidgen (1994) points out, the rich picture is not intended to be an objective representation of the problem situation: '...in preparing a rich picture the analyst is making an interpretation. Consequently there is no single correct rich picture and in one sense a "good" rich picture is one that people recognise as being representative of the situation they find themselves in...'

Stage 3: Root Definitions of relevant systems

The aim is to define notional systems which are relevant to the problem situation. This can be done by choosing an issue or a primary task from the rich picture, then stepping back from the real world and defining a system which addresses that issue or carries out that task. A relevant system must incorporate a particular *Weltanschauung* (i.e., a particular view of the world), which may or may not seem desirable to the definer. Each relevant system will have a Root Definition. Guidelines are provided for checking that a root definition is well formulated, and these are summarised in the mnemonic 'CATWOE'.

Thus a root definition should include:

C Customers who are beneficiaries or victims of the system
A Actors who carry out the defined activities
T Transformations of inputs to outputs
W A *Weltanschauung* (i.e. a view of the world)
O An Owner, who has the power to authorise or dismantle the system
E Environmental Constraints (i.e. elements outside the system which it takes as given).

It is recommended that a variety of root definitions are explored, incorporating different *Weltanschauungen*.

The following is an example of a root definition appropriate for the vehicle rental company:

A rental company owned, staff operated system to meet all of a corporate customer's requirements for staff mobility by supplying appropriate rental vehicles when requested, subject to competition from other rental companies, personal taxation of company car users, and an adequate return on capital employed, in order to secure the survival of the rental company. (Vidgen, 1994)

From this root definition the following CATWOE is derived:

C Corporate customers' management
A VCR corporate services
T Corporate customers' need for staff mobility satisfied
W The survival of the rental company can be secured by supplying
   vehicles to companies that have a need for staff mobility
O VCR management
E Competitors' activity; Inland Revenue taxation policy on company
   cars; an acceptable return on capital employed.

This root definition contains a particular W or view of the world. An alternative view might be that corporate employees prefer to use their own cars, in which case a relevant CATWOE might be:

C Corporate customers' employees (as victim)
A Corporate customers' employees
T Use of private car on company business Æ that use decreased
W Employees without company cars believe that they can subsidise
   their private motoring through expenses claims for business mileage
O Corporate customers' management
E Private car running costs.

The purpose of considering alternative worldviews is to identify how the view might impact upon the choice of primary task model developed for VCR. If the system is developed on the assumption that the W in the first CATWOE prevails when, in practice, corporate employees prefer to use their own cars (the W in the second CATWOE), it is possible that the system will not succeed.

A number of other views (Ws) might be considered: for example, in Vidgen's paper different root definitions that could be relevant to the views of the environmentalist and of the competitors are also considered.

It is important to emphasise that the root definitions and conceptual models developed in SSM are not a reflection of what is, or even of what ought to be. SSM models are better thought of as 'epistemological devices serving coherent discussion' (Checkland, 1995). Thus we develop root definitions and conceptual models that may be more or less relevant to the problem situation as expressed. These models can form the basis for discussion concerning what action might be taken to improve the situation.

Stage 4: Conceptual Models

In this methodology, a conceptual model is a human activity model which rigorously matches a root definition. The activities can be derived from the verbs in the root definition, and the model shows the dependencies between these activities. The inputs and outputs implied by the transformation are also shown. Guidelines are provided for checking that a conceptual model represents a viable human activity system, as defined in the 'Formal System Model': i.e., a human activity system, S, is a 'formal system' (i.e., it is valid) if, and only if:

a. S has an on-going purpose or mission
b. S has a measure of performance
c. S contains a decision-making process
d. S has components which are themselves systems having all the properties of S (i.e., subsystems)
e. S has components which interact such that effects and actions can be transmitted through the system
f. S exists in wider systems and/or environments with which it interacts
g. S has a boundary, separating it from the wider system or environment which is formally defined by the area within which the decision-making process has power to cause action to be taken
h. S has resources, physical and – through human participants – abstract which are at the disposal of the decision-making process
i. S has some guarantee of continuity, i.e., it has some 'long term stability', or will recover stability after some degree of disturbance.

It is also suggested that conceptual models could be validated against other systems thinking.

It usually requires considerable iteration between stages 3 and 4 to produce a matching root definition and conceptual model.

Stage 5: Comparison

The activities in the conceptual model are now compared with what happens in the real world. For each activity, the following types of question are asked:

- Is this activity carried out in the real world?
- How is it done?
- How is performance measured?
- Is the activity carried out effectively?

Stages 6 and 7: Implementing Feasible and Desirable changes

The intention at this stage is to investigate which activities are both culturally feasible and systematically desirable. This may involve: exploring the feasibility of moving from the current situation to the situation implied by the conceptual model; bringing people together to share understanding of their different perceptions of the situation; and getting the people concerned to judge the desirability of the activities. This may lead to the reiteration of Stages 3 and 4.

It is only in the later stages that solutions are considered. The information gathered by the methodology may be used as a basis for designing or choosing a solution, and for planning an acceptable implementation.

These seven stages constitute a basic descripton of SSM, presented to give the reader an overview of the type of issues addressed. In addition to contributing to improving the process of articulating the product concept, problem analysis and the analysis of options, it also:

- provides a way of thinking about the current organisation and identifying the potential for change;
- helps to identify key stakeholders and their objectives;
- helps to identify key workgroups and their objectives;
- helps to identify which work roles should be supported, and why;
- helps to develop descriptions of work roles;
- helps to develop visions and design proposals;
- supports communication between people; and
- helps to identify conflicts between stakeholders, and between viewpoints.

The next section provides an introduction to ETHICS.

3.4 ETHICS

ETHICS (Effective Technical and Human Implementation of Computer-based Systems) is based on socio-technical systems theory (Emery and Trist, 1969). ETHICS was briefly introduced in Chapter One, Section 1.7.5, as a participative approach. The aspects of ETHICS which are key to the problem being addressed in this chapter are that it:

- supports identification of system goals from user, management and organisational viewpoints;
- encourages cooptimisation of the social and technical systems.

ETHICS has some similarities with SSM inasmuch as it involves comparing an ideal situation with the actual situation. In SSM the real world situation as described in the rich picture is compared with the conceptual (systems) model. In ETHICS the work mission (the ideal) is compared with the actual work done and the level of job satisfaction of the staff. Comparison of 'ideal' against 'actual' leads to identification of what needs to be changed, and why.

Mumford describes her work in *Designing Systems for Business Success, the ETHICS Method* (1986), and in Mumford (1989).

The methodology is designed for use by managers and users of new technology, and the following description reflects this. However, it is also used by Requirements Engineers and designers as a technique for problem analysis, because it provides clear guidance of how to conduct the analysis.

According to Mumford, ETHICS has twelve main steps (Mumford, 1986); these are:

1. Specify Work Mission.
2. Describe present work activities and needs.
3. Consider Job Satisfaction.
4. Decide what needs to be changed.
5. Set efficiency, effectiveness and job satisfaction objectives.
6. Consider Organisational Options.
7. Reorganise.
8. Choose Computer System.
9. Train Staff.
10. Redesign Jobs.
11. Implement.
12. Evaluate.

Steps 1 to 5 deal with requirements. However, they do not result in a requirements document being produced. A list of tasks and objectives is identified, and these are used as a basis for choosing a computer system and for reorganisation.

In the description which follows we assume that the manager of a small organisation is applying the methodology to that particular organisation.

Step 1: Specify Work Mission and Identify why change is needed

The manager is asked to specify a personal work mission and the work mission of the business, and is asked to think carefully about the business, the reason for its existence and the things it is trying to achieve.

Next, the manager is asked to state the principle activities required to achieve the work mission, and to examine whether there is a good fit between the principle activities required to achieve the work mission of the business and the activities which are now taking place.

The manager is then asked to examine ways of increasing the possibility of becoming, for example, more efficient, effective and satisfied with work. In essence the manager is asked to identify why there is a need to change the present method of working.

Step 2: Describe Present Work Activities and Needs.

The manager is asked to provide a broad picture of the activities of the business and its staff, as they are at present.

The analysis should describe the following:

1. Day-to-day tasks, indicating which tasks take up most time.
2. The most frequent or more serious work problems that have to be solved.
3. Those aspects of work which require coordination.
4. Those aspects of work where new developments are taking place.

These may be new procedures or new products or services. New ideas being developed should be described.
5. How work is controlled. The kinds of targets that are set and how these are monitored. The most important control procedures should be indicated. (Mumford, 1986)

The manager is then asked to identify: the most important tasks, together with the tasks upon which most time is spent; the most serious problems; where good coordination between activities is required; new methods or ideas that are being developed; the most important control procedures. Tha manager's most important activities, and those of the business as a whole, are also identified.

Step 3: Consider Job Satisfaction

At this stage the manager is asked to examine his or her personal job satisfaction, and that of the staff.

The assumption here (Mumford, 1986) is that, if people enjoy what they are doing, their morale and motivation will be high and they will probably be efficient and effective, as well as satisfied. If however, their morale is low and they experience feelings of frustration, they are unlikely to work at high efficiency and they may derive little pleasure from their jobs.

The job satisfaction questionnaire is used on the manager and on all of the staff in order to ascertain which aspects of their work they particularly like and those which they dislike.

Underlying assumptions of the questionnaire: Job satisfaction is defined as a good fit between what a person does and has in his or her job and what he or she ideally wants. Most people want the following: to use the knowledge which they possess and to increase this; to get a sense of achievement from work; to have access to resources that enable them to work efficiently and effectively; to have an element of personal control so that they can make decisions and make choices; and to have a well designed job that provides the right mix of interest, variety and challenge.

The manager and the staff are asked to complete the questionnaire, to analyse the results and to collate comments on the use of skills etc.: what people like doing most; what people like least; aspects of work the staff are most satisfied with and those they are least satisfied with.

Step 4: Decide what needs to change

Compare the results from Step 1 with those of Steps 2 and 3. Identify which tasks carried out at present are unnecessary and could be removed. Identify which tasks on the list from Step 1 are not actually carried out and should be: i.e., identify any new tasks which will help achieve the work mission.

Having decided what key changes are needed, the manager is then asked to identify the following:

- Changes which could help achieve the business work mission and the manager's personal work mission by improving managerial efficiency.
- Changes which would help the business as a whole to improve its efficiency.
- Changes which would improve the manager's personal effectiveness.
- Changes which would improve the effectiveness of the business as a whole.
- Changes which would improve job satisfaction.
- Future changes.

Step 5: Set Efficiency, Effectiveness and Job Satisfaction Objectives

A clear set of objectives for the manager, the staff and the business as a whole, which are directed towards the achievement of work missions, will enable the manager to:

- Understand exactly what the manager wants to get from any reorganisation of work and new technology, before any changes are made.
- Evaluate the success of any reorganisation of work or new technology once it is introduced, by checking how well it is contributing towards the objectives. (Mumford, 1986)

The manager is asked to set objectives relating to: efficiency, job satisfaction, effectiveness and future change.

Step 6: Consider Organisational Options

Before introducing new technology it is important to ensure that the business is organised and managed in the best possible way, to achieve increased efficiency, greater effectiveness and higher job satisfaction. There is considerable evidence that computer systems introduced into badly organised work situations tend to be failures, whereas computers introduced into well organised situations provide many benefits.

Mumford advises that, if the answer to any of the following questions is 'yes', the manager should consider some reorganisation:

- Would reorganisation help the manager to achieve personal efficiency objectives, or those of the business as a whole?
- Could some of the manager's work problems be eliminated altogether, giving faster and more effective control over the remainder?
- Would reorganisation enable the manager to become more personally effective in critical business areas, and enable the business as a whole to become more effective?

- Would reorganisation remove frustrations an enable improvements in job satisfaction?
- Would reorganisation make the business more flexible and enable it to cope more easily with change in the future?

Step 7: Reorganise: Principles for good organisational design

An incremental approach to change is recommended. The manager is encouraged to take account of the following principles for good organisational design:

a. People work best in groups of six to eight or less.
b. Giving a group responsibility for a part of a business, rather than a single function, increases work interest, responsibility and motivation.
c. Let a group identify and correct its own mistakes, rather than have another group do it. This prevents mistakes being made.
d. Try to ensure that information goes directly to the group that has to act on it. This avoids delay in taking action.
e. Give each group clear work objectives but leave it to them to decide how to achieve these objectives. This encourages responsibility and stimulates initiative.
f. Make sure each group knows exactly what it is responsible for and which other groups it needs to coordinate effectively with.
g. Give each group some development opportunities by, for example, introducing new methods of working or new activities.
h. Involve staff in decisions about what organisational changes to introduce.
i. Keep organisational structure flexible so that it can easily be altered.

Any reorganisation must be linked to the business mission and objectives. (Mumford, 1986)

Step 8: Choose a Computer System

ETHICS does not provide specific guidance on how to choose a computer system. However, at this point it is clear what the objectives of the system are and what tasks it is intended to support. Thus, either an appropriate system design can be developed, or a ready-made package can be purchased and tailored for use.

Step 9: Train Staff

Once a system has been purchased and installed, appropriate training must be given to the staff. At this point, any reorganisation of jobs will occur, leading to Step 10.

Step 10: Redesign Jobs

It is necessary to consider the job of each staff member. Each job should provide the following good design principles:

a. a good fit with the needs of the person doing the job. It should not be so routine as to cause boredom, nor so demanding as to cause stress.
b. Work variety and the opportunity to use a number of different skills.
c. The opportunity to use judgement and made decisions.
d. The opportunity to do a complete job, and see a set of tasks through from start to finish.
e. The opportunity to learn and go on learning.
f. A feeling that the work is important and seen by others as important. (Mumford, 1986)

Step 11: Implementation

Staff participation throughout the process of planning the change is very much encouraged since they are the people who will make implementation a success or a failure. It is important that they 'own' the change.

Step 12: Evaluation

Once the system has settled down, its ability to contribute to the efficiency and effectiveness of the business, to job satisfaction and to the objectives, must be carefully evaluated. Ask staff to complete the job satisfaction questionnaire again and assess whether the objectives set earlier have been achieved. Part of the evaluation will be to identify future change.

In addition to contributing to the process of articulating the product concept and analysing the problem situation from user, manager and organisational perspectives, ETHICS also:

- provides a systematic step by step approach;
- supports communication between people;
- develops knowledge of the current organisation and the potential for change;
- helps identify organisational objectives;
- helps identify work roles to be supported, and why;
- helps describe user characterstics;
- helps identify quality attributes of efficiency, effectiveness and satisfaction.

The next section introduces Eason's approach to User-Centred Design.

3.5 EASON'S APPROACH TO INFORMATION TECHNOLOGY AND ORGANISATIONAL CHANGE

Eason's approach was introduced briefly in Chapter One, Section 1.7.6, as a User-Centred Design (UCD) approach. The aspects of Eason's UCD which are key to the problems being addressed in this chapter are that:

- it supports alignment of business, organisational and technical strategies;
- it supports the construction of system development teams;
- it provides guidance on assessing the costs and benefits of alternative solutions from an organisational and user perspective.

Eason's approach to information technology and organisational change has evolved over a number of years through his work within the HUSAT Research Institute at Loughborough University. His approach is similar to ETHICS in that it is based on socio–technical theory. However, Eason's work is much more comprehensive. The socio–technical approach is supported by a user-centred design process which begins by considering strategic options for the enterprise, and covers all levels of design through to physical options for workstations.

The approach recognises the fact that, for a future system to be successful, it has to have a technical system which is compatible with the social system. Figure 3.4 shows the gap between the two systems.

As Eason points out (Eason, 1992), 'The laws and principles of the two kinds of system are quite different and the methods of constructing them are owned by different specialists. At all stages of development they are likely to be separate and different, perhaps conflicting objectives may be pursued.'

The user-centred design topics are those which reside in the 'gap' referred to in Fig. 3.4, and shown in Fig. 3.5. Another important aspect of

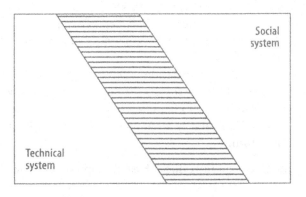

FIG. 3.4 The gap between the two systems.

the approach is the recognition that the 'gap' will only be closed if users and technical specialists work together. Thus consideration of the user-centred design team structure is paramount to the whole process. Finally, no system will be successful unless it meets the business objectives of the organisation. One of the key aims of user-centred systems design is to serve business, organisational and human objectives (see Fig. 3.6).

Eason has summarised the user-centred design process in ten principles. Each principle is supported by specific concepts and methods. It is not feasible to present all the methods or techniques here. Section 3.6 briefly explains one of the techniques, and serves to highlight Eason's approach. A full description of the technique is given in Appendix A.

There follows a brief description of the ten principles. As can be seen in Fig. 3.7, the principles are grouped into four categories (Eason, 1992).

Systems Integration

These are overarching principles which relate to the prime objective of aligning business, organisational and technical strategies:

1. Socio–technical design: The goal is the progressive development of socio–technical systems to serve business objectives.
2. Systems for diverse goals: User-centred design works for the development of integrated solutions which have the flexibility to serve diverse and changing goals.

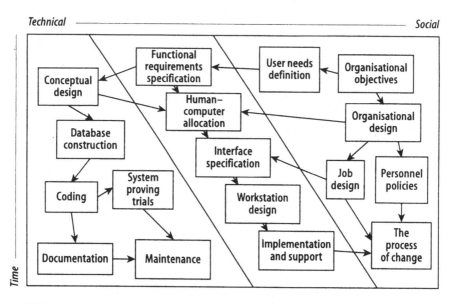

FIG. 3.5 Design topics in the socio-technical systems design of information technology systems (Eason, 1988).

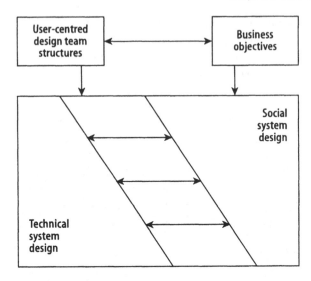

FIG. 3.6 User-centred design of a socio–technical system (Eason, 1992).

User-Centred Design Structures

These are principles for the construction of system development teams which would establish and sustain a partnership between users and technical specialists:

3. Developing user ownership: The design process has to develop a sense of ownership and commitment towards the integrated system among the user community.
4. Involving all stakeholders: Design must be a partnership between users and designers and must seek involvement of all types of stakeholders.
5. Integrating contributions across disciplines: The development of an integrated solution requires the integration of contributions from all relevant disciplines.

User-Centred Design Processes

These are principles by which users and designers can formulate requirements and test alternative solutions:

6. User requirements and values: User-centred design should make explicit user requirements which include the goals and values of the organisation and the users within it, the nature of the tasks to be undertaken and the characteristics of the users.
7. Representing solutions: Informed decisions require the representation throughout design of concrete visions of socio–

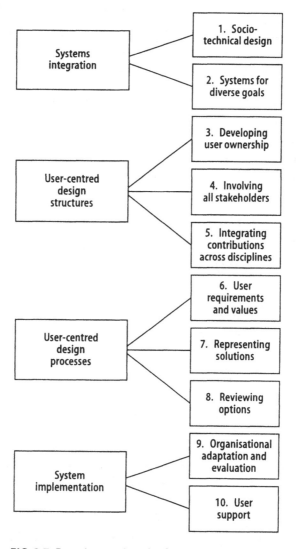

FIG. 3.7 Eason's ten principles for user-centred design.

technical opportunities.

8. Reviewing options: Users need time, resources and structures in which to gather data about the implications of options and to review options against their requirements.

System Implementation

These are principles to govern the manner in which change is undertaken to ensure users maintain mastery over work organisation and the systems they utilise.

9. Organisational adaptation and evaluation: Implementation requires local adaptation through incremental steps and user reviews to provide for evolutionary growth.
10. User support: Implemented systems must contain point-of-need support mechanisms to ensure users can progressively develop the competence to exploit the technological facilities provided.

Eason's ten principles are supported by a large number of techniques. In order to illustrate his approach and to provide the reader with a technique which is useful in avoiding one cause of expectation failure, the next section introduces one particular technique.

3.6 EASON'S TECHNIQUES FOR COST–BENEFIT ASSESSMENT OF THE ORGANISATIONAL IMPACT OF A TECHNICAL SYSTEM PROPOSAL

As indicated above there are many techniques which could be employed in support of a user-centred approach. Only one technique is described here (and in Appendix A). It can be used in support of Principle 6: User Requirements and Values. The technique has been chosen because it supports the early assessment of the organisational impact of a technical system proposal. The five main stages are shown in Fig. 3.8.

The circumstances under which this procedure is most relevant and has been most tested occur when, as a result of a feasibility study, an outline conceptual specification for a technical system has been proposed for a specific organisational setting. A group, perhaps comprising user management, technical staff and user representatives, is charged with assessing this proposal for organisational and user acceptability (Eason, 1988).

The procedure takes a group of this kind through the following stages:

1. *Systems Specification*: stating the technical proposal in a form which facilitates the assessment of organisational impact.
2. *Organisational Specification*: outlining that part of the organisation which will be affected by the system and describing the actual or planned work roles that will be occupied by users of the system.
3. *User Cost–Benefit* Assessment: an assessment of the impact of the system upon the major work roles of potential users.
4. *Organisational Match* Assessment: An assessment of the overall impact upon the organisation.
5. *Socio-technical Design*: a series of checklists to support the development of a strategy for the development of an acceptable socio-technical system.

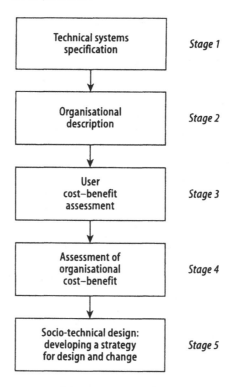

FIG. 3.8 Cost–benefit assessment of the organisational impact of a technical system proposal.

In addition to its value in assessing specific technical proposals, the procedure has also proved useful as a training aid, helping people to appreciate the way in which technical systems influence organisational issues, and thereby developing their ability to play a constructive role in technical developments within their own organisations (Eason, 1988).

The technique includes checklists for: user cost–benefit analysis; organisational cost–benefit analysis; social system design; technical system specification; and the project design process.

Eason states that:

The outcome of completing these checklists should be a broadly-based specification of a socio–technical system which should serve significant enterprise goals and should be functional, usable and acceptable to the user groups within the organisation. Furthermore, the composition of the design process should assign responsibilities for this specification to users so that the system can be developed in detail without losing sight of the user and organisational variables that have been identified as significant. (Eason, 1988)

In addition to contributing to the process of:

- articulation of the product concept;
- problem analysis;
- feasibility studies and cost-benefit analysis of options;

this technique also:

- supports documentation of requirements;
- supports identification of various viewpoints;
- supports reconciliation of viewpoints;
- supports the user in reviewing models developed;
- supports users in analysing their own problems and identifying the need for change;
- supports construction of appropriate requirements teams;
- supports identification of stakeholders;
- supports the development of a 'shared meaning' of the system;
- contributes to visions and design proposals;
- contributes to knowledge of the current organisation and the potential for change.

Appendix A contains Eason's own description of the technique from Eason (1988), and is reproduced with permission.

3.7 SUMMARY

In this chapter, one cause of expectation failure was discussed, the failure to realise that the goals of a system are defined within the total context of an organisation and its political and social environment and not just in relation to technology (Robinson, 1994). A case study illustrated the problem and a number of approaches which would assist the Requirements Engineer in dealing with the problem were presented.

The next chapter is concerned with one cause of process failure and advocates group session approaches as one means of reducing the risk of such failure.

NOTES

1   From Robinson, 1994.

2   References in this format refer to the Inquiry Report of the South West Thames Regional Health Authority, February 1993.

4 Specific Techniques 2: Group Session Approaches

OBJECTIVES

- To address one cause of process failure:
  ...that there is a failure to realise that appropriate human communication mechanisms need to be established as part of the requirements process. If different interest groups do not communicate effectively with each other, each will seek to exert power and influence over the others... (Gasson, 1995; Markus and Bjorn-Anderson 1987)
- To present a case study which illustrates the problem of inappropriate human communication mechanisms.
- To highlight the role of group session approaches in assisting with this problem.
- To discuss the role of the facilitator in group sessions.
- To introduce the key features of Joint Application Design (JAD) sessions and their role in requirements.
- To illustrate the role of Quality Function Deployment (QFD) in requirements.
- To give an overview of Cooperative Requirements Capture (CRC).
- To present some practical techniques for Cooperative Requirements Capture.

4.1 INTRODUCTION

This chapter begins with a case study of a multidisciplinary research and development project. An analysis of the progress of the project is presented. The analysis shows that, although the project began well with agreement on the process model to be adopted, the two main groups drifted apart. Two different requirements documents were produced, one from each group. Neither group developed any real understanding of the needs of the other group, and both groups attempted to take control of the development process. This resulted in eventual failure and abandonment of the project. The analysis explains how and why this happened.

The main problem is identified as a lack of appropriate human communication mechanisms within the project.

The main body of this chapter is dedicated to three different approaches to requirements in which appropriate human communication mechanisms are provided. The three are placed together under the heading of group session approaches, because each is centred around holding a series of group meetings in which the role of facilitator is explicitly defined. The approaches are introduced here because they contribute to requirements twenty-one to twenty-seven from the 'wish list': that is, human communication techniques which:

21. Support construction of appropriate requirements teams.
22. Support identification of stakeholders.
23. Support the development of a 'shared meaning' of the system being specified.
24. Encourage intuition, imagination and common sense among participants.
25. Support communication between people from a diversity of backgrounds.
26. Support facilitated meetings with predefined agendas and problem solving strategies.
27. Support the development of listening skills among participants.

The three approaches were chosen for the following reasons:

- JAD (Joint Application Design) is a well established group session approach in which the role of group members and the facilitator are well defined. In addition, JAD is an example of a structured analysis approach to requirements.
- QFD (Quality Function Deployment) is a group session approach which is gaining recognition within software development circles. The role of the facilitator is well defined. In addition, QFD is an example of a quality approach to requirements.
- CRC (Cooperative Requirements Capture) is a group session approach in which the role of group members and the facilitator are well defined. In addition, CRC is an example of a Human Factors or HCI approach to requirements.

4.2 AN ILLUSTRATIVE PROBLEM SITUATION

The purpose of this section is to present an analysis of a particular project in which two different interest groups were involved. Although the project began with good intentions on both sides, the lack of appropriate communication mechanisms resulted in each group developing a different view of the requirements. This in turn led to attempts to exert power and influence over each other through producing different requirements specifications and following parallel development processes. The two groups never achieved a common understanding of the requirements and, although two prototypes were developed, the project was eventually abandoned.

This analysis is taken from the work of Susan Gasson (1995). Selected extracts are reproduced here with permission.

4.2.1 A Study of a Multi-disciplinary Research and Development Project[1]

This study was concerned with a multidisciplinary research and development project, based at a UK university, to design a computer-based system to support interactive student learning. It was conducted through an analysis of the design documents produced by the project team, and through a series of interviews with project team members. As the interviews took place after a decision had been made to abandon the project, some of the team-members' *post hoc* attitudes could be interpreted as defensive: triangulation[2] was used between interviews, to derive a representative picture of the project. Although an external project sponsor was involved, the sponsor's involvement was limited to contact via progress meetings – for this reason (and as access was complicated by the sponsor's withdrawal from the project), the sponsor's contact staff were not interviewed as part of this study.

From the beginning of the project, there was an explicit recognition of the need for a high degree of user-involvement, to permit evaluation of the student-learning benefits of the system. A decision was made by the project manager, a senior academic psychologist, to recruit equal numbers of psychologists and information systems (IS) professionals onto the team, and to use an iterative prototyping model for the system development process: this model is shown in Fig. 4.2. It is clear that, from the beginning, the psychologists were seen as proxy (and powerful) users by both themselves and the IS professionals on the team: they were there both to evaluate the learning benefits of the target system and to ensure that the system was designed for optimum usability.

4.2.2 The Research Framework[1]

The research framework used for analysis in this study was that proposed by Markus and Bjorn-Andersen (1987), shown in Fig. 4.1. The influence of users in development decisions is constrained by (IS) professionals who may exert power over users in four ways:

- Technical power may be exerted in advocating a particular course of action without providing users with the evidence to make their own evaluations.

- Structural power may be exerted by developing IT policies and practices which constrain user choices.
- Conceptual power may be exerted by shaping users' concepts of what IT can provide.
- Symbolic power may be exerted by shaping user values with respect to IT (normally through the provision of system exemplars).

		Target of power exercise	
		Issues of fact	Issues of values
Context of power exercise	Specific development project	Technical	Conceptual
	IS management policy	Structural	Symbolic

FIG. 4.1 Types of power exercise (Markus and Bjorn-Andersen, 1987).

4.2.3 Research Findings[1]

A comparison of the intended process-model (Fig. 4.2) with the actual process-model (Fig.4.3) of the project is illuminating. From the beginning, there appears to have been a dichotomy of approach between the two disciplines, despite attempts by the project manager to coordinate process-paths, which reflected team-members' disparate interests. Two separate system requirements documents were produced, one reflecting innovations in the use of the system, another reflecting its basis in leading-edge *technology*. Even when the results of the initial requirements documents were combined, two rival requirements specification documents were produced, each reflecting only part of the other perspective.

It would appear that team-members from neither discipline fully understood the requirements of the other discipline and both sub-groups attempted to resolve the resulting cognitive dissonance by prioritising their own requirements. The need for IS professionals and users to learn from each other during system design and development is a common thread in information systems literature: Eason (1982) highlights the time-lag between developer understanding of technical potential and user understanding, while Curtis *et al.* (1988) discuss the critical role of the 'expert designer' – who has prior experience of a particular application-domain – in educating other, technical team-members. However, this team lacked the integrative mechanisms necessary for such learning. Both disciplines attempted to control the development process: the psychologists by agreeing project task-structures and deadlines with the project-sponsor, the IS professionals by using the problematic nature of the unproven technology to separate the technical development processes from learning-evaluation.

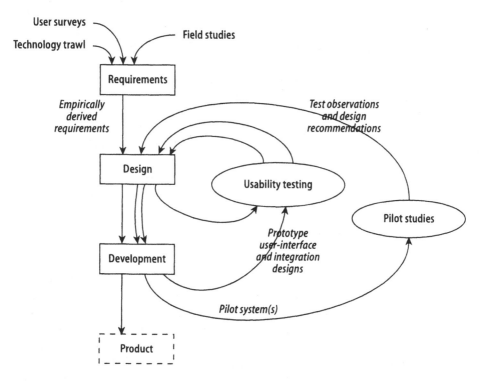

FIG. 4.2 The proposed process model for the R & D project (Gasson, 1995).

The integrated design and development processes from the intended process-model (Fig. 4.2) became split into two separate process-loops, controlled by the two separate halves of the project-team in the actual process-model (Fig. 4.3). In response to the psychologists' attempt to exert structural power (by defining project tasks), the IS professionals gained control of the process by using structural dependencies between the tasks. The technical nature of the production of prototypes for evaluation gave the IS professionals the ability to exert technical power, as the psychologists did not have the expertise to produce these prototypes. Although there was a concerted effort on the part of the psychologists to participate in the design of the initial prototype (Prototype$_1$ in Fig. 4.3), this appears to have been thwarted by their dependence upon the IS professionals to configure the technology.

There appears to have been an implicit agreement between the two IS professionals working on this stage of the project that the first prototype was not intended to be incorporated into the target design, but was produced as a diversionary tactic, to occupy the psychologists while the IS professionals proceeded with the 'real' design. While this was partly a negative reaction on the part of the IS professionals to what were perceived as unrealistic deadlines for the initial prototype (which had been set by the psychologists, in their attempt to gain control over the project), the IS professionals frequently used the term 'flower arrangers' to refer to the psychologists on the team – a revealing metaphor for their perception of the relative value of the contribution of technical and user requirements. When asked explicitly why the design and lessons learned from the first prototype were not used for the second prototype, the response from one of the IS professionals was:

> Well, the cycle broke down because it was such a naff prototype. I think we just generally ignored any requirements that came out [from the psychologists], because we had much better ideas that we felt were ready to go: what we wanted to do for the first 'real' prototype. Obviously our minus one [Prototype$_1$] was produced – but we generally just disregarded it.

The use of the name 'minus one' for Prototype$_1$ reveals its perceived lack of relevance for the intended system outcome: at the same time as the psychologists were evaluating this prototype, the IS professionals were engaged upon the development and evaluation of a prototype for a completely different system design (Prototype$_2$ in Fig. 4.3). This was not communicated to the psychologists. Thus the IS professionals were able to exert symbolic power, by shaping psychologists' expectations of the system: the psychologists were more likely to accept design suggestions from the IS professionals following evaluation of the first prototype, as anything had to be better than the existing design!

The psychologists attempted to exert conceptual power over the IS professionals by the performance of field studies on commercially-available systems for similar purposes. However, these were not read by the IS professionals, who exerted their own conceptual power by prioritising technical requirements over user requirements when selecting appropriate technology. They were able to do this as the psychologists had been placed in a weak position structurally: the evaluation report from the first prototype was not completed, as it became clear at this point that the design of the first prototype had been abandoned. The evaluation results (and, by association, the psychologists' contribution to the project so far) were therefore meaningless. Once a second system prototype had been produced, technical and usability evaluation still took place as two disparate processes, conducted in isolation from each other, with neither

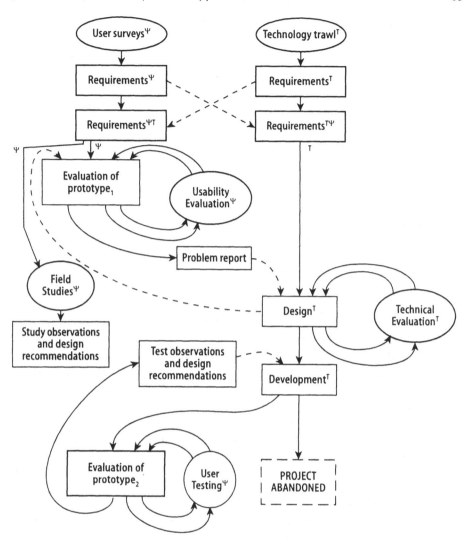

FIG. 4.3 The actual process model for the R & D project (Gasson, 1995). Ψ = activities performed by psychologists; T = activities performed exclusively by computer (technical) professionals; ΨT = activity performed by psychologists, with input from computer professionals; TΨ = activity performed by computer professionals, with input from psychologists

process informing the other. It would appear that the project was then abandoned, for a number of complex reasons – not least that communications between team members had almost completely broken down. Although the ostensible reason given was the withdrawal of the project sponsor, the project failed to find another sponsor because the body of work produced proved insufficiently coherent to attract further funding.

One of the most striking issues which arose from an analysis of the interviews was the formality of communication between the two 'sides' of the project team: the psychologists and the IS professionals. The team size was relatively small – at no time did the core team exceed six members – yet most of the communication between IS professionals and psychologists appears to have been via formal specification documents. It appeared that the IS professionals treated the psychologists on the team as proxy users and therefore dismissed them as an unnecessary distraction from the core task of designing the technical system, while the psychologists felt frustrated and resentful at their dependence upon the technical expertise of the IS professionals: both sides tried to legitimise their priorities by the production of 'official' project documents – an explicit attempt to exert conceptual power. In pursuing reasons for the integrative failure of the two disciplines, it was observed that the team members had been accommodated in two separate offices: one for the psychologists and one for the IS professionals. When asked why the disciplines had not been mixed in their accommodation, the internal project manager commented that the two disciplines had refused to share an office, but the psychologists' perception was that the IS professionals had refused to share an office as they did not want to 'waste' time in educating the psychologists in the technology to be used. The psychologists made repeated requests for tuition in the technology, but these were refused on the basis that meeting them would divert the IS professionals from the tasks necessary to meet project deadlines – another exercise of structural power.

This case study revealingly illustrates the exercise of all four types of power on the part of the IS professionals, most of which appeared to be explicitly directed to exclude the proxy users who were perceived as an unnecessary diversion from the 'real' processes of design. The psychologists were observed to have attempted to exert both conceptual and structural power over the IS professionals, but were unable to exert technical or symbolic power, as they were in a position of dependency with respect to the technical expertise which was required to exert these two types of power in a context where the design process was defined as primarily technical. Even when the psychologists had exerted structural power by taking control of the project deadlines, the IS professionals were able to subvert this control by the use of their technical power, in producing a throw-away prototype.

Also revealed is the importance of recognising both IS developers' and users' learning processes as legitimate design activities. Both interest groups on this project appeared to act as they did because they had insufficient understanding of the requirements and the necessary activities of the other group. If the project process-model had included tasks designed to educate each of the two groups in the others' domain, it is likely that the project would have been more successful. However, integrative communication mechanisms and organisational structures were also needed, to allow the two groups to reach a common set of interests. As the IS professionals were permitted to work together, as a separate organisational unit, and as system design was perceived by the whole project team as being primarily a technical activity, the IS professionals were able to re-define and control the central processes of the project and to exclude the evaluation and use studies which had formed the *raison d'être* of the research project. A comparison of the intended process model with the actual process model (Figs 4.2 and 4.3) illustrates to what extent the IS professionals were able to exclude users from the central processes of design and decision-making.

4.2.4 Conclusions from the Illustrative Problem

A number of points can be drawn from this case study in terms of the themes of this book. First, a process failure can occur even though a project may start with an agreed process model. Second, a process model is useless unless the human communication mechanisms are put in place to support the project team in following the process. Third, although the two groups had a shared objective in the development of the proposed system, they did not share a vision of what the system should do, and why.

The next two sections introduce some of the characteristics of group session approaches and details of the role of the facilitator. In essence the process model will provide the agenda for the group session or, more likely, for a series of agendas for a series of group sessions. Each group session would normally be preceded by some preparatory work and would be followed by some further work. The role of the facilitator is seen as crucial to enabling effective communication between different interest groups.

4.3 CHARACTERISTICS OF GROUP SESSION APPROACHES

There is an increasing recognition of the importance of group session approaches to requirements capture and analysis. Group session approaches to requirements normally involve a number of people coming together in a meeting situation. As with any group meeting, there needs to be an agenda and someone to act a chairperson or facilitator of the meeting. For the group session to be successful people must listen effectively, feel free to express ideas, and be empowered to participate in the process and the outcome of the meeting. The session will be judged to be successful if there is an agreed and documented outcome.

In this section an attempt is made to identify a number of general characteristics of group sessions.

Sharing of workspace

Group sessions usually take place in rooms where the furniture can be arranged to facilitate face-to-face communication between group members. Often the tables and chairs are set in a horseshoe shape so that all members can see all other members, and also see the facilitator. The lighting, heating and general ambience of the room must allow for intense and uninterrupted periods of discussion.

Communication between group members

Group session meetings provide a rich medium of communication. A great deal can be achieved in a setting in which people can see each other and can be sensitive to each other's behaviour and reactions. Someone seen tapping their fingers on the table may be clearly annoyed, someone yawning may have lost interest in the discussion, another person leaning forward and pointing may be agitated or trying to make a point forcefully.

Sharing of information

Sharing of information is essential, both to prevent unnecessary duplication of effort and to ensure that all members can access the same information. The group needs facilities to support the documentation, navigation and retrieval of that information. This is often acheived through the use of whiteboards and flipcharts. For example, results of brainstorming sessions are often documented on sheets of flipchart paper and subsequently pinned up around the room for group members to view and to retrieve the information as required.

Coordination and control of shared objects

In a group session there must be only one version of the object under discussion. For example, consider a requirements team attempting to develop a task hierarchy diagram. A number of versions of the diagram can accumulate and become difficult to manage; more than one person may be modifying the diagram at the same time and there may be a proliferation of associated notes, papers and diagrams which become difficult to maintain.

Decision making

Central to the requirements group is the ability of the group to reach a decision. The decision may be concerned with the objectives of the common task of the group, the method of working to be adopted by the group and the choice of group members.

Specific decision-making techniques may be employed: for example, social judgement analysis, the Delphi technique and the Nominal Group technique (see Viller, 1991b, for further details).

Organisation and common understanding of the work process

In group session approaches the method or technique used largely determines the agenda. It general the group will need to agree on the role of each individual, set specific objectives and deadlines, and decide upon some way of keeping itself informed as to how each person, and the group, is progressing. The facilitator plays a key role in this.

Facilitation

The facilitator plays a key role in helping the group to reach decisions, in managing the shared workspace by deciding who will write on the whiteboard or other shared facility, and in facilitating communication between team members. However, it is not only the trained facilitator who facilitates in group sessions; members of the team themselves may choose to facilitate 'from the floor'. The role of the facilitator is discussed in more detail in the next section.

4.4 THE ROLE OF THE FACILITATOR

This section draws largely on work by Viller (1991a and b).

A feature common to many group session approaches to Requirements Engineering is the notion of someone whose role is to assist the process of group working. That person is referred to generically as a facilitator. The term facilitator itself denotes a set of skills and behaviours that may be applied by, for example, a group-worker, teacher, manager, therapist or a coordinator. The application of these skills may be different in the various contexts. Nevertheless, 'facilitator' implies a readily identifiable, common 'core' of skills and behaviours that may be used by any of the above.

The Shorter Oxford English Dictionary defines the verb 'to facilitate' as 'to render easier; to promote, help forward'. The role of a group facilitator, therefore, is concerned with assisting the other group members in performing their collective task as a group.

At the initial stages in a group's lifecycle, the relationship towards the facilitator may be all that is common to the other group members, and thus the facilitator becomes the group's central person. A facilitator, with his or her knowledge of the group process, can utilise this position to improve group cohesion, and to set the group norms. As a group develops, individuals will begin to identify themselves as members of the group, and the common relationship of everyone towards the central person will become less important (Douglas, 1970). During these middle stages of the group's lifecycle, the central person's role is much more that of enabler, sitting back from the group and only intervening when necessary. Finally, as the group nears the end of its function, the role of the central person becomes more important again, as he or she assists the other members through the process of winding-up the group. The precise role played by the facilitator at this stage will depend upon the circumstances in which the group is breaking up: for example, upon whether or not the group has fully achieved its purpose (Douglas, 1970).

Opinions differ on the facilitator's status within a group. Some of this difference can be explained by the 'bias' of the source. For example, if the facilitator is to perform some leadership function for the group – as in management situations – then he or she will be in a position of power over the other group members. Conversely, if he or she is someone who is brought in from outside as a professional facilitator, then his or her function will be more of an assistant to the group, helping the other group members to achieve their objectives without having any stake in the outcome. This second example describes the facilitator's role in its generic sense, the key factor being that the facilitator is concerned with enabling the process of the group achieving its aims, while having no stake in the content of these aims.

While the dynamic aspect of group work is one of its advantages, problems can develop, and when they occur the facilitator's role takes on greater importance. It is necessary for any facilitator to be able to recognise when a problem is developing, and also to have the skill and knowledge to enable the group to deal with it.

Any action that a facilitator takes to 'correct' group process problems is known as an intervention. Five 'Generic Problem Syndromes' (along with their symptoms, possible causes, and possible interventions) have been identified by Westley and Waters (1988), and are presented in Table 4.1.

A facilitator will usually have at his or her disposal a number of techniques for assisting a group in the decision-making process. These may vary from simple 'brainstorming' to more complex 'structured group process' method (see Viller, 1991b, for further details).

The Requirements Engineering techniques which follow all provide some level of guidance on the role of the facilitator. The group session approaches described below are JAD, QFD and CRC.

4.5 JOINT APPLICATION DESIGN (JAD)

JAD was introduced briefly in Chapter One, Section 1.7.4, as a structured analysis approach. IBM's Joint Application Design (JAD) (August, 1991) draws users and information systems professionals together to design systems jointly in facilitated group sessions. Gibson and Jackson (1987)

TABLE 4.1 Generic Problem Syndromes (after Viller, 1991b).

'Multi-Headed Beast' syndrome

SYMPTOMS	Digressions; interruptions; multiple topics; no listening; no integration of ideas.
POSSIBLE CAUSES	No agreement on agenda; no process design; mixing problem-solving strategies.
POSSIBLE INTERVENTIONS	• Suggest 'round robin' to clarify task • List perceptions of task • Seek synthesis (rephrase, find continuities, categories) • Formulate/reformulate agenda

'Feuding Factions' syndrome

SYMPTOMS	Repetitious arguments; open attacks, anger.
POSSIBLE CAUSES	Hidden agendas/power struggles; fear of change.
POSSIBLE INTERVENTIONS	• Stop action: 'we're having difficulty agreeing on a solution... • Allow individual to list criteria privately • List criteria independently of alternatives • Measure alternatives against criteria.

'Dominant Species' syndrome

SYMPTOMS	'Plops'; 'unequal air-time'; passive/aggressive body language; withdrawal
POSSIBLE CAUSES	Dominance: not heard, frustrated Withdrawn: afraid, frustrated, insulated
POSSIBLE INTERVENTIONS	Direct: question/poll under-participators; thank/limit over-participators Interpretative: At end of meeting, share perceptions on levels of participation • self rating • 'round robin' on views • solicit norms on participation

'Recycling' syndrome

SYMPTOMS	'Broken record' behaviour; irritation with lack of progress; failure to gain consensus.
POSSIBLE CAUSES	Ideas not being recorded; confusion about problem-solving process.
POSSIBLE INTERVENTIONS	• Introduce/reintroduce problem-solving steps • Identify which issues belong to which steps • Identify 'where we are, where we've been, where we're going'.

'Sleeping Meeting' syndrome

SYMPTOMS	Long silences; absence of energy/ideas; withdrawal.
POSSIBLE CAUSES	Fear of volatile issue; hostility; depression, fatigue.

POSSIBLE INTERVENTIONS	• Describe observation – 'blocked condition of meeting'
	• Suggest mood-check
	• Then:
	  • take a break
	  • address underlying problem
	  • decide on action plan to rectify
	• and/or
	  • return to task, allotting time to address the problem at end of meeting.

---

claimed that JAD studies report 20% to 60% increases in productivity over traditional design methods; and further, that JAD promotes cooperation, understanding, and teamwork among the various user groups and information systems staff. JAD defines six different roles which should be represented at a group session. These are: the session leader, the user representative, a specialist, an analyst, an information systems representative and an executive sponsor. JAD teams are given guidance and proformas which can be used as a basis for the agenda for group sessions. However, teams are encouraged to customise these according to their problem situation.

The JAD facilitator is referred to as a session leader. According to Crawford (1994), the session leader manages the process, and facilitates debate and preparation of documents. Within the group session the role of the facilitator is similar to that described in Section 4.4; however, Crawford describes a role which also includes the facilitator taking explicit responsibility for activities outside the meeting situation. The session leader is expected to liaise with the JAD sponsor, to reach agreement as to who should attend meetings, to be responsible for agreeing the agenda with participants, to agree on allocation of work to participants between meetings and to ensure that all appropriate documents and presentations are prepared on schedule. Crawford (1994) gives advice on how to run a JAD session and how to become an effective session leader. Crawford also provides a number sample workbook pages for specific applications of JAD.

The JAD approach can be used at various levels of detail, starting with the business vision and concept analysis through to requirements analysis and specification of designs. JAD activities would be needed for each level of detail. August (1991) describes typical JAD activities as consisting of a JAD/Plan followed by two JAD/Designs. Each activity is composed of three phases: customisation, session and wrap-up.

A JAD/Plan session usually lasts for between one and five days, depending on the size and complexity of the system. The session leader guides participants through eight tasks (August, 1991):

- conduct orientation;
- define high-level requirements (including objectives, anticipated benefits, strategic and future considerations, assumption and constraints, security, audit and control requirements);
- bound system scope (including business flow diagram, system users and locations, out-of-scope functional areas);
- identify and estimate JAD/Designs;
- identify JAD/Design participants;
- schedule JAD/Designs;
- deal with document issues and considerations;
- conclude session.

The JAD/Design session usually lasts for between three and ten days. The session leader guides participants through the following tasks (August, 1991):

- conduct orientation;
- review and refine JAD/Plan requirements and scope;
- develop workflow diagram;
- develop workflow description;
- identify system data groups and functions;
- specify processing requirements;
- deal with document issues and considerations;
- conclude session phase.

The JAD sessions are accompanied by workbooks which contain a collection of proformas for the teams to complete, either during the session or as part of the follow-up activity. Examples of proformas include: participant matrix forms, issues forms, estimating assumptions forms, screen layout forms, report layout forms, interface description forms and function description forms.

In addition to providing support for a group session approach to requirements, JAD also contributes to the techniques 'wish list' in the following way:

For the process of requirements it:
- supports articulation of the product concept;
- supports problem analysis;
- supports feasibility studies and cost-benefit analyses of options;
- supports analysis and modelling;
- supports documentation of requirements;
- supports a systematic step by step approach;
- provides standardised ways of describing workproducts;
- provides procedures for maintaining workproducts.

In terms of human communication it:
- supports identification of various viewpoints;
- supports reconciliation of viewpoints;
- supports the user in reviewing models developed;
- supports users in analysing their own problems and identifying the need for change.

JAD helps to develop knowledge of:
- relevant structures on the users' present work;
- visions and design proposals;
- overviews of technological options.

4.6 QUALITY FUNCTION DEPLOYMENT (QFD)

Quality Function Deployment was introduced in Chapter One, Section 1.7.8, as a quality approach to requirements. QFD (Sullivan, 1986) originated in the Japanese car industry as a means of translating customer requirements into appropriate technical requirements throughout the development and production of a product. QFD is based on group sessions in which the 'House of Quality' (see Fig. 4.4) is used as a focus of attention. The House of Quality is centred around a matrix which shows the relationship between the customer requirements and the proposed product features.

In QFD the customer requirements provide a central theme and are used as a basis for setting targets for the design and implementation. Traditional QFD is split into four iterative phases: product planning, parts deployment, process and control planning, and production plan-

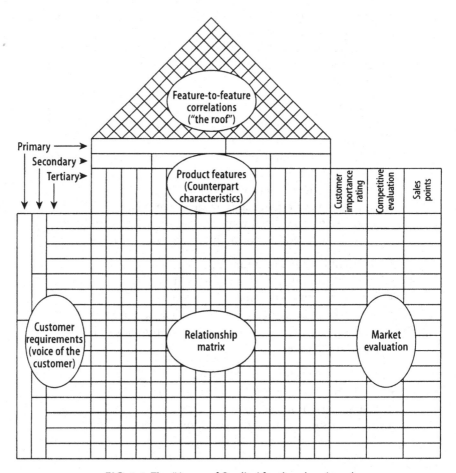

FIG. 4.4 The 'House of Quality' for the planning phase.

ning. All four phases are similar: each has its own House of Quality and associated group sessions. The people who attend the group sessions will be those who are responsible for a particular phase of the product and who need to come to an agreement concerning their actions within that phase.

Betts (1989) encourages people to apply QFD to software projects. Figure 4.5 illustrates how she sees QFD fitting into the software development lifecycle.

The customer voice provides the driving force for identifying the measurable objectives for the product. The objectives are then used to drive the high level design, and so on, in such a way that the customer voice acts as a driver for the whole software development process.

Zultner (1989) provides more detailed guidance on how to use QFD on a software project. He illustrates how to incorporate dataflow diagrams and entity relation diagrams into the design phase. In Zultner (1993) he demonstrates how a wide range of tools and techniques can be linked to QFD in order to develop a Total Quality Management approach to software development. A key theme of the paper is that of process improvement through project teams identifying and setting their own targets while at the same time being aligned to the organisation's vision for improvement (Zultner, 1993).

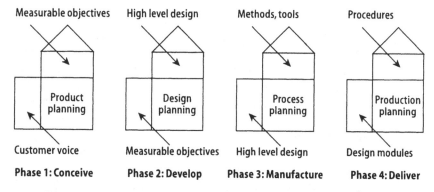

FIG. 4.5 QFD and the software development lifecycle (after Betts, 1989).

Reports of the use of QFD in various parts of the computer industry (Cohen, 1988; Daetz, 1989) have claimed reductions of 17% in product definition time, and clear traceability of requirements from initial design through to full production.

Marsh (1991) describes facilitation of the team effort as critical for the successful completion of a QFD study. Proper facilitation, he claims, will help the QFD team fully integrate the talents, skills and creative potential of each team member. The facilitator has a coordinator's role in the planning, design, execution and completion of a QFD project. The facilitator

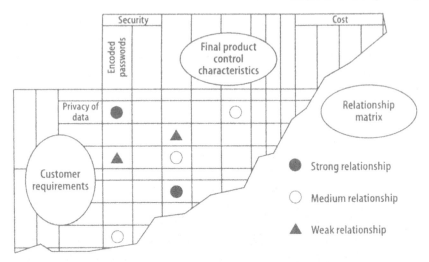

FIG. 4.6 Step 3 of the planning phase.

is described as a neutral, non-evaluative and non-manipulative group focal point. Marsh has produced an excellent guide, called 'Facilitating and training in Quality Function Deployment', which gives details of the roles and responsibilities of both the facilitator and the project manager of a QFD project (Marsh, 1991).

The remainder of this section describes the activities associated with phase 1, the planning phase, of QFD. The outcome of the planning phase is the overall customer requirement Planning Matrix, as shown in Fig. 4.5. This translates the voice of the customer into counterpart control characteristics: that is, it provides a way of turning general customer requirements (drawn from market evaluations, comparison with competition, and market plans) into final product control characteristics (Sullivan, 1986).

Putting the planning matrix together involves eight steps:

Step 1: The first step is to state the product requirements in customer terms. These are entered into the left hand column of the matrix in Fig. 4.5 under the heading 'customer requirements'. The requirements may be grouped together into primary, secondary and tertiary requirements. Primary requirements are often quite general, e.g., 'easy to learn'; secondary and tertiary requirements would be more specific.

The information needed for Step 1 usually comes from a variety of sources. No specific guidance is given as part of the method, despite the fact that this is obviously a critical step in the whole process.

Step 2: Step 2 is to list the 'product features' across the top row of the matrix. These are the final product characteristics which will be needed

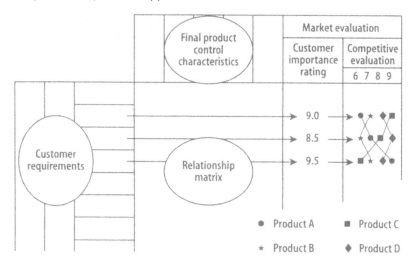

FIG. 4.7 Step 4 of the planning phase.

in order to meet the stated customer requirements. The characterisics must be capable of being expressed in measurable terms.

Step 3: Step 3 is to develop the 'relationship matrix'. This shows the relationship between the customer requirements and the product features (the final product control characteristics). The relationship is expressed as being strong, medium or weak (see Fig. 4.6). The benefit of filling out this relationship matrix using appropriate symbols is that it quickly demonstrates whether the final product control characteristics adequately cover the customer requirements. Absence of symbols (or a majority of 'weak relationship' signs) indicates that some customer requirements are not addressed. At this point, product characterisics may need to be modified or supplemented to ensure that all customer requirements are adequately addressed (Sullivan, 1986).

Step 4: Step 4 is to add the market evaluation. This shows the customer's rating of the importance of each customer requirement and an evaluation of how well the competition is doing in meeting each requirement (see Fig. 4.7).

Step 5: Step 5 is a comparison of the final product control charactistics against the performance of the competition. This step involves identifying the numerical, measurable performance rating, for example, of competitor products A, B, C and D, against each of the product characteristics. This gives a measure of the level of achievement of the competition.

Step 6: Step 6 is concerned with using the analysis from Step 5 and the customer importance rating from Step 4 to identify the selling points for the proposed product. These points might consist of statements such as 'Best-in-class at meeting customer requirement X', or 'Lowest energy consumption components for meeting customer requirement Y'.

Step 7: Step 7 now requires the QFD team to identify the actual measurable targets (control characteristic targets) which must be achieved. These targets are based on the agreed selling points, the customer importance rating, and current product strengths and weaknesses.

Step 8: Step 8 in the development of the Planning Matrix involves the selection of product control characteristics that are to be deployed through the remainder of the QFD process.

In addition to providing a group session approach to requirements, QFD[3] also contributes to the following requirements from the 'wish list':

- supports articulation of the product concept;
- helps develop visions and design proposals;
- supports identification of requirements for generic products;
- is capable of working alongside market analysis techniques;
- supports analysis of competitive products;
- supports predictions of future users and future use, and estimations of future usage;
- supports generic descriptions of typical users and groups of users;
- supports identification and specification of quality attributes.

4.7 COOPERATIVE REQUIREMENTS CAPTURE (CRC)

The approach developed by the present author through a number of collaborative projects is called Cooperative Requirements Capture (Macaulay, 1993). This is a group session approach, similar to JAD in that the role of participants and the role of the facilitator are clearly defined. Participants consist not only of users and designers but also include those with a stake in the system being proposed. Stakeholders are defined as all those who have a stake in the change being considered, those who stand to gain from it, and those who stand to lose. The CRC facilitator does not have a stake in the proposed system, but does have knowledge of the method and possesses the interpersonal skills required of a facilitator.

Some of the stakeholders identified have a direct responsibility for the design and development of the various system components, and hence have a major interest in being involved in the requirements capture process. Others have a financial responsibility for the success of the system and therefore may also need to be involved. Yet others, who will be the recipients of the resulting system, also have a major contribution to make in terms of specific task knowledge and the ability to assess the likely effects of the new system. The choice of representative stakeholders to attend the group sessions is of vital importance to the success of the session.

For example, in a requirements analysis for a proposed theatre box office system, the following stakeholders were involved: from the theatre: the accountant, the box office manager, an experienced member of the box office staff, and the theatre manager; from the software house: the project manager, the senior designer, the designer and the user support person. Figure 4.8 illustrates the role of the CRC group session in bringing the two organisations together.

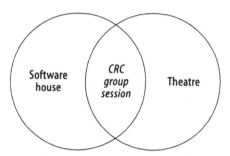

FIG. 4.8 CRC group session: scenario 1.

In contrast to this, Macaulay (1994) reports a case study of the use of CRC to determine the information requirements of electricity control room engineers in the year 2000. In this case three separate organisations were involved in the group session: one was the research and development unit associated with the electricity supply industry, and the other two were separate electricity distribution companies. There were four types of stakeholders involved:

i. Strategic thinkers' from all three organisations; these were people with a long-term view of their company's future direction.
ii. Computer specialists from all three organisations, who would ultimately be responsible for the design and development of the proposed system.
iii. Control room engineers from the two distribution companies who would be representative of the users of the proposed system.
iv. Managers of the control rooms who would ultimately be responsible for the introduction of the proposed new system.

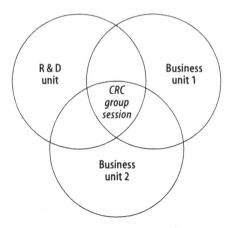

FIG. 4.9 CRC group session: scenario 2.

The role of the CRC group session is highlighted in Fig. 4.9. CRC is different from JAD and QFD in that the focus of attention in CRC is the user. The CRC method is as follows:

i.   Identify the problem.
ii. Formulate the team.
iii. Group session 1: explore the user environment.
iv. Validate with users.
v.   Group session 2: identify the scope of the proposed system.
vi. Validate with stakeholders.

Each group session has a number of steps: for example, session 1 includes: the business case; workgroups; users; tasks; objects; interactions; and consolidation. Each step includes an introduction, brainstorming, prioritisation and generation of agreed descriptions, using checklists and proformas which deal with user related issues. For a fuller description and a detailed case study of group session 1, see Macaulay (1995b).

In the CRC approach, the user and the user environment provide the focus of attention for the stakeholders, and help them to develop a shared vision of the future system. They 'explore' the user environment together, and they are encouraged to describe what users do now and to develop a vision of how things might change in the future. They develop a shared understanding of the potential for change and a shared terminology for discussing the problem domain. Figure 4.10 gives an overview of the CRC method.

The rectangular shaped boxes represent face-to-face meetings or workshops. Normally the meeting is supported by a trained facilitator, who guides the team through the main steps of the method and who also encourages all the stakeholders to participate. The oval shaped boxes

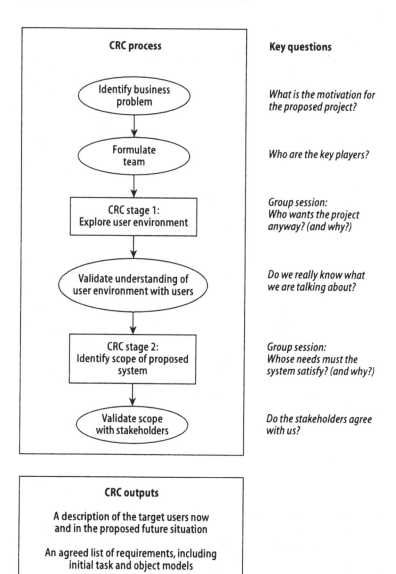

FIG. 4.10 CRC: overview of the process.

represent activities which must take place either before or after a workshop. These activities usually involve some subset of the stakeholders
consulting with others outside the immediate team. Communication and

consultation is important in order to elaborate and validate the requirements before the detailed design begins.

Each part of the approach is described briefly below.

4.7.1 Identify Business Problem

It is assumed that there will be some motivation for proposing that a future system be developed. That motivation may come from some specific business need: for example, the need to improve the speed of customer service at the check-in desk at an airport, or the need to provide bank customers with easy access to cash withdrawals. In some cases it may be that the commissioning organisation is planning ahead and that the business problem may refer to some future need: for example, electricity distribution companies may ask: 'What kind of computer support will control room engineers need in the year 2005?'. In other cases, it may be that only a small incremental improvement is needed: for example, an estate agent may already have a computer-based system but may want to improve the quality of the information stored.

If the commissioning organisation is a software house or computer company it may be that they are in possession of some innovative technology and they wish to identify whether target users do have a need for that technology, and to be able to describe what that need is.

In all these cases, a business need is identified, whether it be an improvement in customer service, a future need, a small incremental improvement to the existing system or a need to use some technology which is available. There are many different motivations for proposing that a future system be developed. What is important to the application of the CRC approach is that the business need is articulated and described in such a way that it can be shared with the other team members.

4.7.2 Formulate Team

This will normally involve the project manager or project initiator and the facilitator in the identification of stakeholders and hence of the requirements capture team. Ideally between six and nine stakeholder representatives will participate in the CRC process. The team would be drawn from each of the four categories of stakeholders identified earlier in the chapter.

4.7.3 CRC Stage 1: Explore User Environment

Exploring the users environment means that the requirements capture team must collectively investigate the target users' organisational setting and identify and describe what they do. The term 'explore' is used because the team is encouraged to 'find out' afresh, to share knowledge about users

TABLE 4.2 User Document: list of contents.

---

1. Management Summary
(including the business case and a brief description of the
proposed system)

2. Organisation/Workgroups
    2.1 Workgroup Control Sheets
    2.2 Organisation Chart
    2.3 Workgroup Table
    2.4 Workgroup Description Checklists

3. Generic Users
    3.1 Generic Users' Control Sheets
    3.2 Generic Users' Description Checklists

4. Tasks
    4.1 Task Control Sheets
    4.2 Task Hierarchy
    4.3 Task Description Checklists

5. Objects
    5.1 Objects Control Sheets
    5.2 Object Structures
    5.3 Object Description Checklists

6. Interactions
    6.1 User/Task/Object Interactions
    6.2 Initial List of Requirements and Attributes

7. Consolidation
    7.1 Statement of Credibility
    7.2 Further Investigations Needed

8. Worth Proceeding?
    8.1 User/Stakeholder Perspective
    8.2 Business Perspective
    8.3 Plan of Action

9. Conclusion

---

and to set aside preconceptions about what users need. They also assess the likely costs and benefits of change from the users' point of view and produce a document recording the shared view of the users' environment.

The structure of the 'User Document' is shown in Table 4.2.

The User Document is produced by the Cooperative Requirements Capture team as a result of a two-day face-to-face meeting.

4.7.4 Validate Understanding of User Environment with Users

Representative users do participate in the workshop but other users will need to be consulted or interviewed in order to ensure that the team has reliable information about all those users who may be affected by the system. After Stage 1 of CRC, but before Stage 2, validation of the information recorded in the User Document should occur and it should be expanded or updated where necessary.

The techniques used for validation will depend on the specific problem: for a generic product further market research may be needed, for a bespoke system specific user interviews may be necessary. In any case, the extent of the information gathering task will depend on the extent of the knowledge and expertise of the stakeholders who take part in the workshop. A team who is highly conversant with their users may need to do very little validation. It is important to note that at this point in the development process highly detailed knowledge of all the users' tasks may not be necessary. The team needs to have enough reliable information to be able to decide which users, which tasks and which objects need to be computer supported and to decide what the extent of that support should be: that is, they need enough information to decide the scope of the proposed system.

4.7.5 CRC Stage 2: Identify the Scope of the Proposed System

This also involves a two-day face-to-face meeting of the requirements capture team and the use of proformas and checklists. At this stage the scope of the system is discussed. The scope of the proposed system is determined at a number of levels.

First, the stakeholders decide which work roles are to be affected and then, for each work role, they decide what the role of the system should be in supporting that role. In particular, the role of the system is decided in terms of the extent of task sharing and degree of control and monitoring of tasks. This is discussed more fully in Macaulay, 1995b, under the heading 'Allocation of Function'. The likely acceptability of this proposed change is considered. In addition, for each work role identified, an initial task model is produced. This helps to clarify and consolidate the understanding of the team with respect to specific roles.

Second, the team is asked to consider which objects from the user environment – i.e., those contained in the User Document – are likely to be of interest to the system: that is, which objects will the system need to hold information about, which will it need to interact with, and which will remain entirely in the user domain.

The scope of the proposed system is determined by the extent of support for the work roles and by the list of objects the system will need to support. In addition, the scope of the system is reviewed from the point of view of each of the major stakeholders to identify whether their needs will be met. The list of requirements also is reviewed from the viewpoint of each stakeholder.

Once the scope of the system is decided and the list of requirements reviewed, the team is asked to identify and agree on usability targets for the proposed system. The process involves matching the current situation with the proposed situation and specifying those areas where usability issues may be a problem. For example, the 'gap' between possessed and required knowledge to undertake a role or a specific task within a role may highlight a potential usability problem. The usability is specified in five parts: the user (*who*); the specified activity (*doing what*); the location (*where*); the targeted performance (*ideal, worst* and *best cases*); and measuring instrument (e.g., *a benchmark*). These parts are all recorded in a usability specification table. In addition, notes should be made about when in the development cycle the usability evaluation should take place: e.g., early, at the prototyping stage, or late, at the installation stage. It is important that usabilty does not only address the end-users but also considers the facilitating users (e.g., installers and maintenance engineers).

The outcome from this stage is an 'Initial Requirements Document' containing an agreed set of requirements for the proposed system. In addition the CRC team will have gained a thorough understanding of the users.

The Initial Requirements Document is the major output from Stage 2 of CRC. Each part is aimed at a specific set of stakeholders and covers a distinct set of issues. Table 4.3 shows a list of contents.

TABLE 4.3 Initial Requirements Document: list of contents.

---

1. Management Summary
(including the business case and a brief description of the proposed system)

2. The Human Requirements
    2.1 Description of the objectives of the commissioning organisation
    2.2 List of the stakeholders together with their objectives
    2.3 List of key workgroups and users and their objectives

3. The High Level Functional Requirements
    3.1 List of work roles to be supported and why
    3.2 Description of each work role in terms of users, objects and tasks

4. The Detailed Functional Requirements
    4.1 Consolidated list of objects to be supported
    4.2 Descriptions of each object together with details of user tasks associated with each object

5. The Quality Attributes
    Quality attributes may include usability, reliability, portability, performance, security, maintainability and acceptability, depending on the proposed system.

6. The Organisation and User Assistance Requirements
    6.1 Documentation requirements
    6.2 Training requirements
    6.3 User support
    6.4 Human–computer interface requirements

7. The Technological Requirements and Constraints
    7.1 Known hardware requirements (user or supplier)
    7.2 Known software constraints (user or supplier)

This Initial Requirements Document is produced in draft form as part of CRC Stage 2. The document represents a statement of the scope of the proposed system, which will then need to be validated by reviewing its contents with the stakeholders.

4.7.6 Validate Scope with Stakeholders

The scope of the proposed system should be validated. A range of techniques could be used at this point, depending on the type of system under consideration. Appropriate techniques might include: use of questionnaires; interviewing users; building mock-ups; throw-away prototypes; or holding focus groups.

Once the scope of the system has been agreed and documented, resources, timescales, tasks, milestones and deliverables can be evolved and the software designer can then proceed with the detailed design.

In addition to providing a group session approach to requirements, CRC also contributes to the following from the 'wish list':

- supports articulation of the product concept;
- supports documentation of requirements;
- helps develop relevant structures on the users' present work;
- helps develop visions and design proposals;
- supports identification of requirements for generic products;
- is capable of working alongside market analysis techniques;
- supports generic descriptions of typical users and groups of users;
- supports identification and description of current work practices;
- supports identification of constraints such as cost, time and security;
- supports identification and specification of acceptance criteria
- supports identification of organisational objectives, of key stakeholders and their objectives, and of key workgroups and their objectives;
- supports identification of work roles to be supported and why, and descriptions of each work role, and functional requirements to support each work role;
- supports identification and specification of quality attributes: usability,;
- supports identification and specification of requirements for user documentation, requirements for training, requirements for user support;
- supports identification and description of human–computer interface requirements.

Appendix B contains a 'user guide' to CRC Stage 1.

4.8 SUMMARY

In this chapter, one cause of process failure was discussed. The case study illustrated the problems caused by inadequate human communications mechanisms within requirements. Group session approaches were introduced as a means of improving human communication within requirements.

The next chapter focuses on one cause of interaction failure, the lack of understanding of users' work.

NOTES

1 From Gasson, 1995.

2 Triangulation is a process of verifying the results from more than one perspective.

3 Sources of additional information: for overviews of QFD see Sullivan, 1986; King, 1989. For guidance on how to conduct QFD see Zultner, 1989; Bossert, 1991; Marsh, 1991. For case studies see Cohen, 1988; Lecuyer, 1989.

5 Specific Techniques 3: Interactive Approaches

OBJECTIVES

- To address one cause of interaction failure:
  ...that designers[1] do not not fully understand the work of users.
  (Greenbaum and Kyng, 1991)
- To introduce techniques which encourage designers to interact with users.
- To discuss the notion of designers as apprentices to users.
- To introduce future workshops and metaphorical design.
- To distinguish between prototyping and cooperative prototyping.
- To introduce Cooperative Evaluation as a Requirements Engineering technique.

5.1 INTRODUCTION

The aim of this chapter is to introduce techniques which will assist the Requirements Engineer (RE)* in understanding the work of users. The techniques introduced here have a common theme of 'interaction': that is, they cause the Requirements Engineer to interact with the users. This approach is in contrast to more traditional 'requirements elicitation' techniques where the RE has a predefined model of what he or she is looking for. For example, using structured analysis techniques, the RE studies the work of the user in order to find processes, data, inputs and outputs; using an object-oriented approach the RE is looking for objects, services and messages; or using a task analysis approach the RE is looking for user tasks, task hierarchies and actions on objects. Using other techniques such as interviewing or issuing questionnaires also relies on the RE having some pre-defined notion of what he or she wants to find.

The problem with these traditional approaches is that the RE may only

---

* In previous chapters, the abbreviation RE denoted 'Requirements Engineering'.

see the work of the user in terms of some predefined model and fails to see much of what the user is actually doing. The work of ethnographers has raised the awareness of the Requirements Engineering community of the shortcomings of existing models of the work of users.

Ethnography is a method derived from anthropology and is based on observing the behaviour of groups. Professional ethnographers are employed to observe users over a long period of time and to make detailed observations about work practices. Analysis of audio-visual recordings and results from field studies 'reveal the delicate and complex web of interactional practices through which information is communicated and tasks accomplished ... even apparently individual tasks like reading, writing or typing into a computer are embedded in interactions with others and are designed in relation to another's activities.' (Luff *et al*, 1993).

The studies of Luff *et al.* suggest that the traditional understanding of tasks as activities carried out by individuals is flawed and that all work activities involve some level of social interaction. By presenting a different model with which to view the work of the users, the ethnographers have pointed to weaknesses in our present models.[2]

The failure to understand the importance of social interaction stems from a general weakness of not understanding the intricacies of what people do and how or why they do it. Requirements Engineers typically have some predefined notion of what they are looking for and do not develop a sufficiently intimate knowledge and understanding of users to see clearly what is actually happening.

Thus the problem faced by Requirements Engineers is how to develop relationships with users to a sufficient level of intimacy to understand the intricacies of their work and to discover how it might be improved.

The next section illustrates some of the problems that arise when the designers of the system fail to understand properly the work of users.

The remainder of the chapter is given over to the description of a number of techniques which were designed to support interaction between the designers and the users. The techniques covered are: Focus Groups; the Designer-as-Apprentice; Future Workshops; Prototyping; Cooperative Prototyping; and Cooperative Evaluation. A complete step-by-step guide to Cooperative Evaluation can be found in Appendix C.

All the techniques described in this chapter contribute to items 28–33 from the 'wish list': that is, they support development of:

28. relevant structures on the users' present work;
29. visions and design proposals;
30. overviews of technological options;
31. concrete experience with the users' present work;
32. concrete experience with the new system;
33. concrete experience with technological options.

5.2 AN ILLUSTRATIVE PROBLEM SITUATION

The purpose of this section is to present a number of examples which illustrate that designers have not fully understood the work of users. The examples are taken from a variety of studies of real-life situations.

The first example illustrates that designers have not thought of things which are obvious to users. Bravo (1993) relates experiences of using a database program:

> If you have ever had to enter data into a database and had to manage it, you know that one of the most common things you have to deal with is duplications. You enter a name and then find that person was already on your list and you want to go back and delete the dup. Say you have two 'Gloria Williams'. If you delete 'Gloria Williams', you have no 'Gloria Williams'. You have to trick the computer: change one of the 'Gloria Williams' to 'Gloria Wilhelm' and then delete 'Gloria Wilhelm' so that 'Gloria Williams' will still be on your list. Why isn't there a simple thing that says: 'dup, delete one'? The computer would know there are two; take out one and you have what you need.

This demonstrates quite clearly that the designers have not studied how or why users delete records from a manual system. In this case not only have they not fully understood the work of users in the first place but neither have they taken the trouble to evaluate the system once it is in use.

The second example, also from Bravo (1993), illustrates that designers have not thought about the effect of the system on the users:

> TWA has a new call distribution system. It has eliminated the six seconds that you used to have between calls to finish scribbling your paper work, or take a sip of water, or maybe crack your neck. Clearly, whoever designed the system had no idea what it would feel like the instant you hang up to have to pick the phone up again.

This demonstrates one of the key weaknesses within Requirements Engineering: that is, the failure to realise that change in the system will cause change in the user's job. In this case there was probably only a minor change in the computer software, but the quality of the user's job deteriorated dramatically.

The third example illustrates that designers have not thought about the effect of the system on social interaction. The example is taken from an ethnographic study of doctor and patient interaction during consultation sessions. The doctor has a computer on the desk and enters patient details into the computer in order to obtain a printed prescription.

...They (the doctor) may also need to attend to output messages, such as requests for clarification or correction of input and/or warning beeps. In addition, once they have started along the prompt line, they will have little control over the order in which information is entered. On occasions, these constraints appear to undermine the doctor's ability to participate simultaneously in discussions with patients concerning topics that are not directly related to the production of prescriptions... (Luff *et al.*, 1994)

This demonstrates that the doctor–patient relationship is being affected by the presence of the computer. The patient can no longer chat to the doctor once the formal consultation is complete because the doctor's attention is drawn towards the computer. Even as the doctor becomes more experienced in using the system, the situation does not change:

However, with one exception, the doctors' increasing familiarity with the system has not as yet led to a marked reduction in the extent to which its use adversely affects social interaction. (Luff *et al.*, 1994)

Thus, not only is the system affecting the way in which the doctor performs his or her job, it is also affecting the relationship between the doctor and the patient.

The following sections present a number of techniques which have been designed to support interaction between designers and users.

5.3 DESIGNER-AS-APPRENTICE

Beyer and Holtzblatt (1995) describe the fundamental problem of interaction between designer and user as one of creating the right relationship between the user and the designer to enable learning. They take the relationship between master and apprentice as a model, in which the user is the master and the designer is the apprentice. The master teaches the apprentice by doing the work and talking about it, while the apprentice learns by watching and listening.

According to Beyer and Holtzblatt, the designer, taking on the role of the apprentice, 'automatically adopts the humility, inquisitiveness, and attention to detail needed to collect good data. The apprentice role discourages designers from asking questions in the abstract and focuses them on ongoing work.' (Beyer and Holtzblatt, 1995)

However, the user–designer relationship differs from the master–apprentice relationship in two important ways:

• The designer must understand and be able to articulate the structure of the work, the constraints on getting the work done, how work is allocated between users and the physical environment in which the work is carried out.

- The designer must think of ways of improving the work and ask users for feedback on proposals for change.

The apprenticeship model provides the designer with an opportunity to explore the work of the user beyond the initial assumptions. It encourages a more intimate relationship in which the designer can develop an empathy with the intricacies of the user's job.

Designer-as-apprentice can:

- support articulation of the product concept;
- support the user in reviewing models developed;
- support users in analysing their own problems and identifying the need for change;
- support the development of a 'shared meaning' of the system being specified;
- enable designers to get concrete experience with the users' present work;
- support identification of constraints such as cost, time and security;
- support identification and specification of quality attributes: usability, reliability, portability, performance, security, maintainability, acceptability and so on depending on the proposed system;
- support identification and specification of requirements for user documentation, requirements for training, requirements for user support.

Beyer and Holtzblatt (1995) describe a number of examples of 'designer-as-apprentice' in practice.

The next section presents a different approach to designer–user interaction, which involves the designer/Requirements Engineer in organising focus groups for users.

5.4 FOCUS GROUPS

The purpose of a focus group (Draper and Oatley, 1991) is to allow a group of users to talk in a free-form discussion with each other, in the presence of the Requirements Engineer.

Focus groups will give the Requirements Engineer insights into how users think and what things are important to them. It is difficult to achieve this with other techniques.

When a focus group is used for the purpose of understanding user needs and requirements, the Requirements Engineer can act as facilitator. In this case the role of the facilitator is relatively passive.

If a prototype or mock-up of a system exists, then the focus group can take place after the target users have had some exposure to the system. The Requirements Engineer will need an audio recorder and a group ideally numbering between four and six users, and should prepare a list of topics for discussion. The topics should be introduced one at a time, the main aim being to find out what people think about the topics. The Requirements Engineer should not take part in the discussion but should facilitate free discussion between the users. A focus group would typically last about forty-five minutes.

If the users have been carrying out particular tasks on a prototype or mock-up, the topics for discussion could relate to those tasks. For example, Topic 1: How did they get started with the first task?; Topic 2: Were they able to complete the first task? If not, what difficulties did they have? Such open questions should trigger the group to discuss problems and share experiences. The Requirements Engineer can listen to the group at the time and listen to the audio recording after the focus group has finished. Often it is useful to have a transcript of the discussion, though this is very time consuming to prepare.

Focus groups can:

- support articulation of visions and design proposals;
- support articulation of the product concept;
- support users in analysing their own problems and identifying the need for change;
- support the development of a 'shared meaning' of the system being specified.

The next section introduces an approach which requires more interaction between designers and users than the focus group.

5.5 FUTURE WORKSHOPS, METAPHORICAL DESIGN AND DESIGN MOCK-UPS

Future workshops were originally developed by Jungk and Mullert (1987) for citizen groups (in Scandinavia) who wanted a say in the decision-making processes of public planning authorities. Kensing, 1987, was one of the first to propose their use in system development. The technique was meant to 'shed light on a common problematic situation, to generate visions about the future, and to discuss how these visions can be realised' (Kensing and Madsen, 1991). Participants should share the same problematic situation, they should share a desire to change the situation according to their visions, and should share a set of means for that change.

Future workshops are run by one or two facilitators and no more than twenty participants. The facilitator's role is to ensure an equal distribution of speaking time and to ensure that all participants can follow the discussion. The facilitator could be a designer or a Requirements Engineer.

Kensing and Madsen (1991) describe a scenario in which future workshops are used together with metaphorical design. Metaphorical design essentially consists of identifying possible metaphors which could be used to help the participants think about possible alternative future situations. The steps undertaken in the scenario can be summarised as follows:

Setting up
- The facilitators met with the project initiator (in this scenario the project initiator was the chief librarian).
- A project group was formed consisting of the two facilitators and six workers (in this case the six were three librarians and three clerical staff).
- A deadline was set for delivery of the proposal for future computer usage, the project group to deliver to the project initiator (in this case the deadline was four months hence).

Preparation
- The group developed a common background by: (a) the facilitators working as 'apprentices' to the workers, and (b) the facilitators demonstrating hardware and software by arranging visits to other workplaces (in the scenario the other workplaces included libraries, a store, a museum and a community centre).
- The facilitators identified possible metaphors (in this case the metaphors for a library were a warehouse, a store and a meeting place).
- Invitations were sent out for a one-day Future Workshop (in this case invitations were sent to all employees, but not to library heads).
- A suitable location and materials were found (tape, markers, large sheets of paper).

Future Workshop: Introduction
- Introduction to the technique and plan for the day by the facilitators.

Future Workshop: Critique Phase
- *Critique Phase Part 1* This was a structured brainstorming session in which participants were asked to focus on current problems at work. Speaking time was restricted to thirty seconds and short statements were written on the wall charts. Statements did not need to be justified by the speaker, and other participants were not allowed to criticise the statements (in this scenario, typical statements included: 'we never talk to borrowers', 'librarians are just attendants', 'we only provide self-service', 'the library is like a supermarket').
- *Critique Phase Part 2* Facilitators encouraged the participants to

group the short statements under four or five headings, and then to separate into small groups for discussion. This was then followed by a plenary session in which all groups presented the result of their discussion back to the meeting (the headings in this case were: 'the library as a warehouse', 'relation to borrowers', 'the organisation of the library', 'the library as a store', 'the role of DMK [one of the project's initiators]').

Future Workshop: Fantasy Phase
- *Warm-up* Facilitators encouraged participants to draw pictures of the library of the future, and hung these on the wall around the room.
- *Brainstorming* Participants were encouraged to write short statements on the wall charts about the future situation. No criticism of ideas was permitted. A 'utopian outline' was produced by getting participants to rank the statements in order of those they favoured most. Each participant was given five votes (statements were made such as: 'tear down the walls', 'arrange small reading groups', 'enable access to the library from home').
- *Use of metaphors* Participants were encouraged to think of the library as a warehouse – in what way was it the same, and in what way was it different? A view emerged from this discussion and from the 'utopian outline' that, although the library had much in common with a warehouse, it should also be like a meeting place.

Future Workshop: Implementation Phase
- *Discussion* Each group discussed the 'utopian outline' and how it might be realised. The suggestions were coordinated in a plenary session and became the basis of a common strategy for library staff.
- *Follow-up actions* Designers were charged with developing a prototype of part of the proposed system. Users were charged with getting support from other library staff and with developing a set of criteria for evaluating the prototype.

As Kensing and Madsen (1991) point out, the future workshop provides a good basis for further development of the project, and additional workshops may be held to solve specific problems encountered later in the project.

Future workshops can:

- support articulation of the product concept;
- support problem analysis;
- support users in analysing their own problems and identifying the need for change;
- support the development of a 'shared meaning' of the system being specified;

- encourage intuition, imagination and common sense among participants;
- support communication between people from a diversity of backgrounds;
- provide facilitated meetings with predefined agendas and problem solving strategies;
- support the development of listening skills among participants;
- support the development of visions and design proposals.

Examples of the use of future workshops can be found in various reports within Ehn *et al.* (1990), Ehn and Kyng (1991), Greenbaum and Kyng (1991), Kyng (1991) and Schuler and Namioka (1993).

The next section considers the role of prototyping in involving users in design. Traditional approaches to prototyping are contrasted with cooperative prototyping.

5.6 PROTOTYPING

Prototyping is the technique of constructing a partial implementation of a system so that customers, users or developers can learn more about a problem or a solution to that problem (Davis, 1993).

There are a variety of approaches to prototyping, each providing an opportunity for users to gain hands-on experience before the final application is built. The two main approaches are: incremental prototyping, where the prototype incrementally becomes the system; and 'throwaway' prototyping, where the prototype is a 'mock-up' or simulation of some aspect of the system designed to get feedback from users.

Although one of the main reasons for prototyping is to involve users, they are normally involved in evaluating some prototype which has already been designed. Designers start from the users' more-or-less articulated requirements, and develop a prototype which tends to take the perspective of the designers and the software engineers themselves.

Users are not normally involved in the design of the prototype itself. Figure 5.1 illustrates the relationship between users and designers.

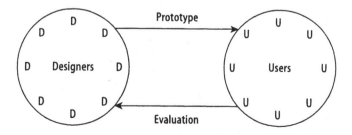

FIG. 5.1 Prototyping: the designers and users belong to separate workgroups.

Traditional prototyping approaches involve the designers in designing the prototype and then getting the users to use it to carry out specific tasks. The results of the user evaluation will then be used by the designers to improve the design of the prototype, or will act as input to the requirements document.

Evaluation can be carried out in a usability laboratory or in the users' normal place of work. In a usability laboratory, video cameras and data logging allow designers or evaluators to view users using the system, and to analyse any problems after the users have completed their tasks. In the users' normal place of work the evaluator would need to observe the user in an unobtrusive manner and make a log of user actions and user problems. Users can be asked to say out loud what they are doing, and this can be recorded on audio tape for later analysis. Further details of evaluation techniques can be found in Macaulay (1995b).

Prototyping used as a requirements technique:

- supports an iterative approach to requirements;
- gives the users concrete experience with the proposed system;
- gives the users concrete experience of technological options;
- helps to identify support for user training;
- helps to identify support for human–computer interface design.

The next section introduces a variation of prototyping which assumes much greater user involvement.

5.7 COOPERATIVE PROTOTYPING

The cooperative prototyping approach advocated by Bødker and Grønbaek (1991) seeks to involve users in the design of the prototype. The aim is to establish a design process where both users and designers are participating actively and creatively. It requires access to flexible, computer-based tools for the rapid development and modification of prototypes. Figure 5.2 serves to illustrate the difference between traditional prototyping and cooperative prototyping.

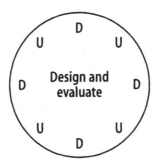

FIG. 5.2
Cooperative prototyping:
designers and users belong to
the same workgroup.

In cooperative prototyping the designers and users belong to the same workgroup, and both participate in the design of the prototype.

Bødker and Grønbaek (1991) view cooperative prototyping as a learning process in which the prototype itself has a secondary role. The process of developing the prototype is of primary importance. The users who belong to the designer–user workgroup learn from the process, and can then educate other, future, users while the final computer system is being developed.

For organisations with a large user population, Pape and Thoreson (1987) suggest a process in which an intermediate prototype is built in one organisational setting. Designers then take that prototype to a second organisational setting and develop it further with a second group of users. Figure 5.3 illustrates this process, and shows one possible role for cooperative prototyping in Requirements Engineering. First, there must an understanding of the overall aim of the proposed system; then, appropriate people must be found to participate in each user group. Each user group may belong to different departments, their needs may be different from each other, but ultimately they must all use the same system.

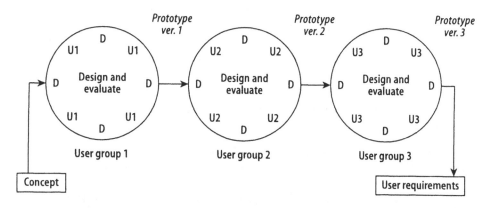

FIG. 5.3 The role of cooperative prototyping in Requirements Engineering.

Designers work with usergroup 1 to develop the first version of the prototype according to their knowledge and understanding. Version 1 of the prototype is evaluated by a second user group who then, together with the designers, develop version 2 of the prototype according to their knowledge and understanding. Each user group learns something of the needs of the other workgroups, and suggests changes according to their own needs. The designers should also learn about the core requirements and variants for each user group.

Cooperative Prototyping contributes to Requirements Engineering in the following way:

- supports communication between designers and users;

- helps to resolve conflicts between different usergroups;
- helps to develop design proposals;
- gives the users concrete experience of the proposed system;
- gives the users concrete experience of the technological options;
- helps to identify key workgroups and their objectives;
- helps to identify which work roles should be supported and why;
- helps to identify requirements for user support, training and human–computer interface design.;
- contributes to an understanding of the quality attributes: usability and acceptability;
- helps to identify core requirements and variants for each usergroup.

Bødker and Grønbaek (1991) describe several case studies in the application of cooperative prototyping. The next section considers an approach which involves designers in user evaluations of prototypes.

5.8 COOPERATIVE EVALUATION

Cooperative Evaluation (Monk *et al.*, 1993) is a technique to improve a user interface specification by detecting possible usability problems in an early prototype or partial simulation. Cooperative Evaluation was developed at York University as an inexpensive means of improving the user interface. It involves users in design by having them complete tasks set by designers.

Monk *et al.* (1993) claim that the distinctive features of Cooperative Evaluation are that:

- it is practical – it can be carried out by designers with very little training and without expensive facilities such as 'usability laboratories';
- it is cost effective, in that it reveals important usability problems in a relatively short time;
- it is for use with early designs and prototypes that may not be complete;
- it brings together designer and user in a cooperative context; the user completes work using the prototype and is encouraged to think aloud about the problems encountered.

Cooperative Evaluation has four main steps, shown in Fig. 5.4.

In the first step, the designer has to identify the target user population and to recruit some typical users. The next step is to identify and prepare some tasks which are representative of the tasks users will do and which will enable the users to explore the features of the prototype. The third step involves the designer in preparing the evaluation by deciding what needs to be done before the users arrive, when they arrive (but before

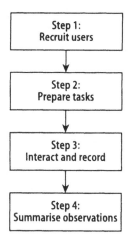

FIG. 5.4
Cooperative evaluation has four
main steps.

starting the tasks), while they are using the system, and during the debriefing session. This step involves the designer and the users actually carrying out the evaluation. Finally the designer must make a summary of observations, mainly in terms of unexpected behaviour and user comments.

The outcome of the whole process will be the development of concrete experience of users using the prototype, a report of observations made, and a list of suggestions for improving the user interface specification.

Cooperative Evaluation contributes to Requirements Engineering by:

- supporting an interactive approach to user interface specification;
- providing a systematic, step-by-step approach;
- supporting communication between users and designers;
- providing the designer with concrete experience of users using the prototype;
- contributing to providing the user with concrete experience of technological options;
- providing a basis for identifying usability quality attributes.

Appendix C provides a step-by-step user guide to Cooperative Evaluation. This is reproduced with permission from Monk *et al.* (1993).

5.9 SUMMARY

This chapter presented a number of techniques for encouraging designers to interact with users. A lack of understanding of users' work is seen as a key cause of low system usage: i.e., interaction failure. The techniques covered were: Focus Groups, the Designer-as-Apprentice; Future Workshops; Prototyping; Cooperative Prototyping; and Cooperative Evaluation.

The next chapter enables the reader to identify the circumstances under which the techniques discussed in Chapters Three, Four and Five might be used.

NOTES

1   In this chapter, the terms 'designer' and 'Requirements Engineer' are used interchangeably. Both refer to the person who defines what the system will do.

2   Ethnography is not a requirements technique, nor is it normally appropriate for an RE to carry out an ethnographic study. However, the results of ethnography can be used to inform the requirements investigation (see Sommerville *et al.*, 1993, for a fuller explanation).

6 Requirements and the Customer–Supplier Relationship

OBJECTIVES

- To introduce a number of alternative relationships between the customer and the supplier.[1]
- To introduce a number of alternative models of the Requirements Engineering process.
- To identify the characteristics of the customer–supplier relationship and of the Requirements Engineering process.
- To discuss how the Requirements Engineer might develop a portfolio of requirements techniques.
- To consider the relationship between the contents of the portfolio, the customer–supplier relationship and the Requirements Engineering process.
- To conclude this text.

6.1 INTRODUCTION

A portfolio is described in the Penguin English Dictionary as a 'large flat case for carrying papers, drawings etc'. A sales representative may carry a portfolio of product descriptions. Depending upon the needs of a potential client the appropriate product descriptions may be drawn from the portfolio and discussed with the client. The portfolio may contain a large number of product descriptions but, in a given situation, the sales representative may use only four or five.

It is in this sense that the term 'portfolio' is used here. A Requirements Engineer (and associated co-workers) may need to have a large number of Requirements Engineering techniques in their portfolio. Appropriate techniques will be drawn upon depending on the given situation.

The purpose of this chapter is to discuss the basis upon which a Requirements Engineer might develop a portfolio of techniques.

The candidate techniques for the portfolio are limited to those techniques described in Chapters Three, Four and Five. Many more techniques could be drawn upon from many different sources, as discussed in Section 1.7. However, in this chapter, only the techniques covered in this book are considered, in order that the reader might see typical situations in which they are used. Thus the candidate techniques are: SSM, ETHICS, Eason's cost–benefit assessment, JAD, QFD, CRC, Designer-as-Apprentice, Focus Groups, Future Workshops, Prototyping, Cooperative Prototyping, and Cooperative Evaluation.

The underlying assumption of the discussion within this chapter is that not all techniques will be needed for every project in which the Requirements Engineer is involved.

Thus, the discussion which follows centres around the question: 'If all the techniques are not needed, which techniques are needed, and when?'

In the sections which follow, an attempt is made to describe a number of Requirements Engineering scenarios. Each scenario has two major components: first, a description of the customer–supplier relationship; and second, a model of an associated Requirements Engineering process. The thesis is that the techniques needed in the portfolio will depend upon the characteristics of both components.

Figure 6.1 illustrates the assumption being made throughout the chapter that the nature of the customer–supplier relationship will determine the model of the Requirements Engineering process adopted, and that this, in turn, will determine the contents of the portfolio.

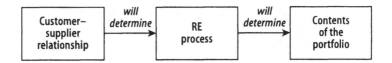

FIG. 6.1 Key influences on the contents of the portfolio.

Seven scenarios are considered in total. In the first four, the customer and the supplier belong to different organisations.

1. A customer issues an Invitation To Tender to a number of potential suppliers.
2. A specific supplier is asked directly to respond to a specific customer request.
3. A supplier wishes to make a generic product which will meet the needs of a (large) number of customers.
4. A supplier has a generic product which needs to be tailored to meet a specific customer need.

In the final three scenarios, the customer and the supplier belong to the same organisation. In this case the customer–supplier relationship depends on the internal organisational structure.

5. The business (customer) and the Information Technology (IT)

function (supplier) are completely separate, and operate as individual businesses.

6. The business (customer) is functionally separate from the IT function but each IT sub-function has clearly defined responsibilities for aspects of the business.
7. The IT function is integrated within the business function, with each business unit having its own IT staff.

Each scenario is now considered in turn. For each scenario a typical Requirements Engineering task is described. This is followed by a brief description of the customer-supplier relationship and an associated model of the RE process. The characteristics of each scenario are highlighted and candidate techniques for the portfolio are suggested.

Note that feedback loops and iterations are deliberately not shown in the process models which follow but, obviously, they do exist. At this level of detail, the author simply wishes to demonstrate that there are a number of different models of the RE process.

6.2 SCENARIO ONE: AN INVITATION TO TENDER (ITT)

A customer issues an Invitation To Tender to potential suppliers.

A typical example of the scenario

The Department of Social Security issues an ITT to a number of major business system suppliers for a new front office system for social security counter clerks.

Customer–supplier relationship

The customer-supplier relationship is shown in Fig. 6.2. Here, the same ITT is issued to a number of suppliers. The suppliers rely on the written document for their source of understanding of what is needed.

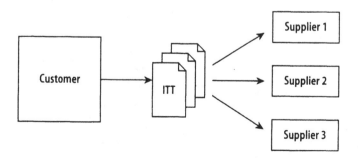

FIG. 6.2 Scenario One: a customer issues an invitation to tender (ITT).

A typical RE process

The ITT is a formal document and, once it is received by the supplier, a contracts team will be put together to decide whether to respond to it. This team is made up of people who are stakeholders and who, as such, have an interest in seeing the project completed. The team members will be from different areas of the company and their task will be to put together a description of the business opportunity from the supplier's point of view.

The business opportunity will be taken up by the 'bid team' who will write a proposal. The proposal is presented to the customer in order to convince them that the supplier can produce the required product. A proposal could be in the form of a prototype, or of a slide show describing the finished product. If the customer accepts the proposal, a contract is signed. This contract is the actual requirements statement which contains details of what should be delivered to the customer. Figure 6.3 gives an overview of the process.

At the next stage, the 'solution team' may be put together. The solution team will use the requirements statement and convert it to a functional

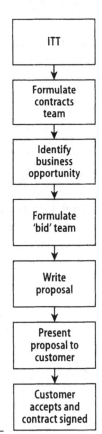

FIG. 6.3
A model of the RE process for
Scenario One.

requirements specification that is in a suitable form for the designers to use. At this stage the customer–supplier relationship changes, and a more detailed requirements investigation may be possible. The characteristics of this stage become similar to those in Scenario Two.

Characteristics of Scenario One

- It is competitive.
- There is limited knowledge of the customer.
- There is probably no access to potential users.
- The ITT represents the customer's own assessment of the requirements.
- The process is limited by time, cost and formal contracts.
- It is strategic for the supplier.
- Many decision-makers may be involved.
- The requirements reside within a formal contract.

Candidate RE techniques for the portfolio

1. Prototyping for demonstration purposes.
2. Design mock-ups.
3. Competitor analysis (aspects of QFD).
4. Cooperative requirements capture workshops for the bid team.

6.3 SCENARIO TWO: RESPONDING TO A SPECIFIC CUSTOMER REQUEST

A specific supplier is asked directly to respond to a specific customer request.

A typical example of the scenario

A supplier is asked by an international vendor of paper packaging to develop a system which will help to reduce stock levels to a minimum.

The customer–supplier relationship

Figure 6.4 shows that, in this case, there is only one customer and one supplier, and that there is a two-way communication between them. The customer may wish the supplier to assist in the development of a written statement of requirements. The supplier may want a statement of requirements to form the basis of a contract with the customer. Requirements statements should be developed to be in the form: 'The system shall do *xyz*'. This will enable the supplier to demonstrate that all the requirements have been met.

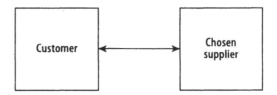

FIG. 6.4 Scenario Two: a specific customer is asked to respond to a specific customer request.

A typical RE process

First, a team of senior stakeholders from both organisations formulates a requirements team. The stakeholders would ideally include those responsible for design and development, those responsible for user support, user managers, user representatives, those with a financial interest and those with a management and strategic interest in the proposed system (the team members should be empowered as decision makers). This team is responsible for agreeing upon which problems need to be solved, for agreeing on the scope of the proposed system, for exploring alternative solutions, for assessing the feasibility of the proposed change and for 'signing' the requirements document. Figure 6.5 shows a typical model of the RE process for Scenario Two.

Figure 6.6 presents an alternative process model for Scenario Two, in which an iterative approach is adopted. In this case a team from the supplier may act as Requirements Engineers, successively identifying customer requirements through a number of iterations. The first iteration starts with the customer's stated goal for the system. The supplier identifies the appropriate client set within the customer organisation, develops a system description, identifies objectives and constraints and considers alternative routes to meeting the objectives, develops a model of the requirements in whatever form is appropriate for the client set, and finally evaluates the model with the client. The result of the evaluation may be the identification of further goals for the system, in which case another iteration occurs.

Characteristics of this scenario

- There is relatively easy access to customers and users.
- The statement of requirements evolves through a series of system descriptions.
- Success is measured in terms of meeting organisational objectives.

Candidate RE techniques for the portfolio

1. Soft Systems Methodology.

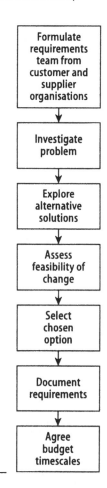

Formulate
requirements
team from
customer and
supplier
organisations

↓

Investigate
problem

↓

Explore
alternative
solutions

↓

Assess
feasibility of
change

↓

Select
chosen
option

↓

Document
requirements

↓

Agree
budget
timescales

FIG. 6.5
A model of the RE process for
Scenario Two.

2. JAD.
3. CRC.
4. Eason's cost–benefit assessment of organisational change.

6.4 SCENARIO THREE: DEVELOPING A GENERIC PRODUCT

A supplier wishes to make a generic product which will meet the needs of a (large) number of customers.

A typical example of the scenario

A large IT supplier wishes to develop a new generation of point-of-sale terminals for use by department stores, supermarkets and other retail outlets.

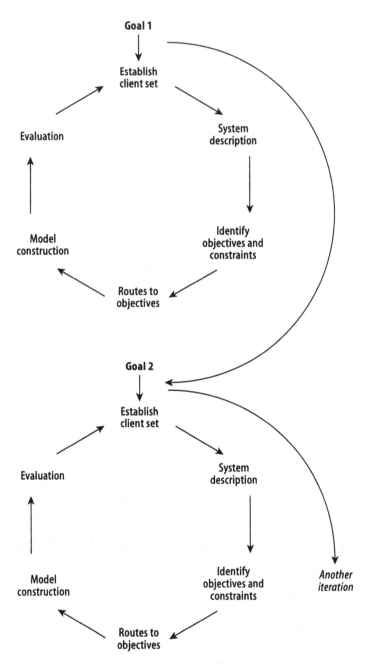

FIG. 6.6 An alternative process model for Scenario Two.

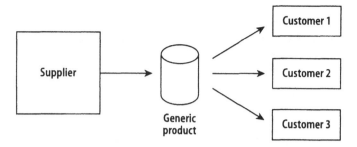

FIG. 6.7 Scenario Three: a supplier wishes to make a generic product which will meet the needs of a (large) number of customers.

The customer–supplier relationship

Figure 6.7 shows a typical customer–supplier relationship for Scenario Three. The supplier wishes to develop a product which will meet the needs of many different customers. Some customer requirements will be generic to all customers; others will be specific to certain types of customer.

In this scenario; requirements exist at a number of different levels. At the first level, market needs must be considered. Questions may include: 'Is the market for the proposed product a mature market?'; 'Are there certain features that the market expects fom this product?'; 'Is there still potential for a substantial market share?'; 'What are competitor products offering?'; and 'How will the market develop in five to ten years from now?' Once a potential market opportunity has been identified and the associated requirements documented, the needs of customers can be considered. Figure 6.8 illustrates that there are different levels of requirements.

Of course, these 'levels' are not independent of each other: for example, identification of some future user requirement may lead to a reassessment of market needs.

A typical RE process

Figure 6.9 illustrates a typical RE process for Scenario Three.

Characteristics of this scenario

- It is difficult to gain access to customer organisations.
- It is difficult to gain access to users.
- It is difficult to gain access to competitors' future plans.
- It involves many decision-makers.
- It is strategic for the supplier.
- There is potentially high risk.
- There is interplay between many factors.

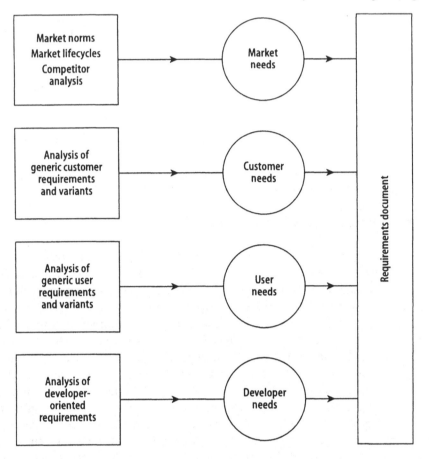

FIG. 6.8 Sources of requirements for Scenario Three.

- Requirements exist at many different levels.
- Requirements reside in many different documents.
- Success is measured in terms of market share.

Candidate RE techniques for the portfolio

1. CRC.
2. QFD.
3. Eason's cost–benefit assessment of organisational impact.
4. Focus groups.

6.5 SCENARIO FOUR: TAILORING A PRODUCT

A supplier has a generic product which needs to be tailored to meet a specific customer need.

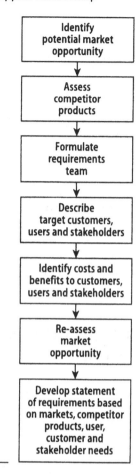

FIG. 6.9
A model of the RE process for
Scenario Three.

A typical example of the scenario

An engineering company wishes to install a stock inventory system. The supplier already has a generic stock inventory system, and will need to tailor it to the specific needs of the engineering company.

The customer–supplier relationship

The customer–supplier relationship for this scenario is shown in Fig. 6.10

Characteristics of this scenario

- The customer has an idea which needs to be articulated.
- There is easy access to the customer.
- Options are limited because of the existing product.
- Success is measured in terms of low cost, user satisfaction and short delivery time.

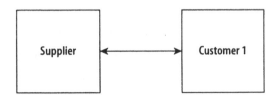

FIG. 6.10 Scenario Four: a supplier tailors a generic product to meet a specific customer need.

Candidate RE techniques for the portfolio

1. Focus Groups, to find out what customers/users are thinking.
2. Cooperative Evaluation, to enable the customer to evaluate the shortcomings of the generic product and to articulate their own needs.

6.6 SCENARIO FIVE: RESPONDING TO A BUSINESS CENTRE WITHIN THE SAME ORGANISATION

The business (customer) and the Information Technology (IT) function (supplier) are completely separate, and operate as individual businesses, but belong to the same organisation.

A typical example of the scenario

The Personnel Department requests amendments to the Personnel System in order to meet new government regulations.

The customer–supplier relationship

The customer–supplier relationship for this scenario is shown in Fig. 6.11.

Characteristics of this scenario

- Communication between customer and supplier occurs at the level of the business.
- It is most likely to be concerned with enhancements to existing systems rather than the development of new systems.
- Success is measured in terms of user satisfaction.
- Requirements are communicated through documented procedures.

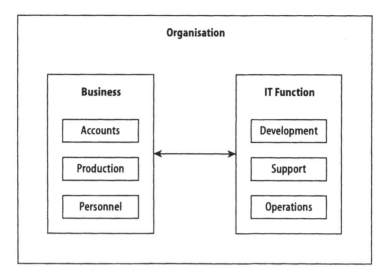

FIG. 6.11 Scenario Five: The business and IT functions operate as separate businesses within the same organisation.

- A service level agreement probably exists between customer and supplier.
- Users are difficult to access because of internal organisational boundaries.

Candidate RE technique for the portfolio

1. Focus Groups.

6.7 SCENARIO SIX: RESPONDING TO THE NEEDS OF A SPECIFIC BUSINESS FUNCTION

The business (customer) is functionally separate from the IT function but each IT sub-function has clearly defined responsibilities for aspects of the business, and belongs to the same organisation.

A typical example of the scenario

The Accounts Department manager requests a new system which will support some future internal restructuring of his department.

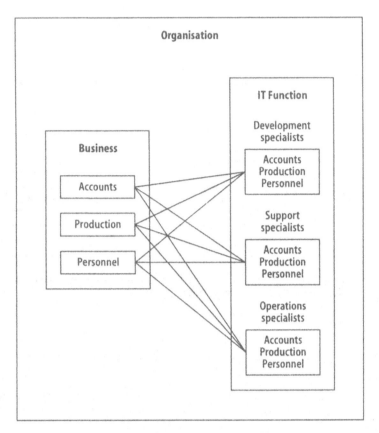

FIG. 6.12 Scenario Six: The business and IT functions communicate at the level of sub-function.

The customer–supplier relationship

The customer–supplier relationship for this scenario is shown in Fig. 6.12.

A typical RE process

Figure 6.13 shows an RE process which is typical of this scenario.

Characteristics of this scenario

- Customer and supplier communicate at the level of sub-function.
- There are good working relationships between customer and supplier.
- The requirements document can be evolved jointly by the customer and the supplier.
- Success in measured in terms of meeting organisational objectives.

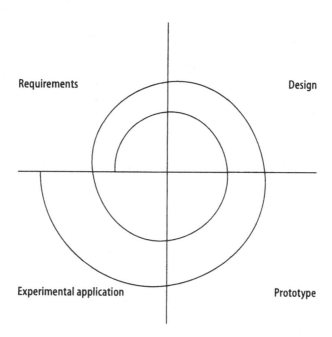

FIG. 6.13 A spiral model of the RE process for Scenario Six.

Candidate RE techniques for the portfolio

1. Future Workshops.
2. Cooperative Prototyping.
3. Cooperative Evaluation.

6.8 SCENARIO SEVEN: RESPONDING TO THE NEEDS OF COLLEAGUES WITHIN THE SAME BUSINESS

The IT function is integrated within the business function, with each business unit having its own IT staff.

A typical example of the scenario

Professionals put together their own set of tools to meet their needs.

The customer–supplier relationship

Professional users consult IT staff directly and take responsibility for their own actions. The customer–supplier relationship is shown in Fig. 6.14.

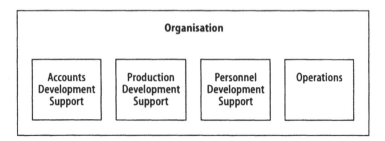

FIG. 6.14 Scenario Seven: the IT function is integrated within each business function.

Characteristics of this scenario

- There is a more intimate relationship, supporting incremental change.
- Success is measured in terms of user satisfaction.
- There are few formal documents.

Candidate RE techniques for the Portfolio

1. Cooperative Evaluation.
2. Designer-as-apprentice.

6.9 CONCLUSIONS

Throughout this book the author has sought to introduce the reader to a number of specific and useful techniques which have not previously been included within mainstream Computer Science literature.

In Chapter One, Requirements Engineering and the task of the Requirements Engineer were introduced. Nine different approaches to the problem of requirements were briefly described. They were marketing, psychology and sociology, object-oriented analysis, structured analysis, participative design, Human Factors and Human–Computer Interaction, Soft Systems, Quality and Formal Computer Science.

In the Introduction to Chapter Two, it was suggested that the objective of the Requirements Engineering process was to specify a system which ultimately proved to be successful. Three types of system failure were introduced from Lyytinen and Hirschheim (1987). They were: process failure; interaction failure; and expectation failure. Requirements Engineering was discussed in terms of these types of system failure and the cause of each type of failure was explored and discussed in some detail. Five possible causes were identified:

1. Lack of a systematic RE process.
2. Poor communication between people.
3. Poor management of people and resources.
4. Lack of appropriate knowledge or shared understanding.
5. Inappropriate, incomplete or inaccurate documentation.

Each cause was discussed in some detail, and the need for various Requirements Engineering techniques was identified. The chapter concluded with a 'wish list' of seventy techniques and a matrix showing a mapping from the 'wish list' to the approaches discussed in Chapter One. This served to reinforce the view that no single approach to requirements will provide the Requirements Engineer with all the tools and techniques needed.

The next three chapters focused on particular areas within the matrix. The areas were chosen to represent particular groupings of techniques which address common causes of system failure. Each chapter dealt with a different type of failure. Chapter Three focused on expectation failure, Chapter Four on process failure and Chapter Five on interaction failure.

In Chapter Three, one cause of expectation falure was examined: the failure to realise that the goals of a system are defined within the total context of an organisation and its political and social environment, and not just in relation to technology. This problem was illustrated through an analysis of the failure of a computer aided despatch system at the London Ambulance Service. The techniques described in this chapter encourage the Requirements Engineer to consider the social, political and organisational issues as part of the requirements investigation. The techniques described belong to Soft Systems, ETHICS and Eason's User Centred System Design. Appendix A contains a 'user guide' to Eason's 'Cost–benefit assessment of the organisational impact of a technical system proposal.'

In Chapter Four, one cause of process failure was examined: that different interest groups do not communicate effectively with each other, each seeking to exert power and influence over the other. The illustrative problem situation was a study of the development of a computer aided learning system, in which two interest groups failed to reach agreement about the requirements despite the fact that two prototypes of the system were developed. The techniques described in this chapter encourage the Requirements Engineer to consider the importance of facilitation in the RE process. They are JAD, QFD and CRC. Appendix B contains a 'user guide' to Stage One of CRC (Cooperative Requirements Capture).

In Chapter Five, one cause of interaction failure was examined: that designers do not fully understand the work of users. A number of typical problems are quoted from a variety of case studies. The techniques described in this chapter encourage the Requirements Engineer to interact with the users. They include Future Workshops, Design Mock-Ups,

Cooperative Prototyping and Cooperative Evaluation. Appendix C contains a 'user guide' to Cooperative Evaluation.

Chapter Six presented a discussion of how the Requirements Engineer might develop a 'portfolio' of requirements techniques based on seven different scenarios. For each scenario the customer–supplier relationship was explained, a typical process model was presented and techniques were suggested for the portfolio.

This book has only covered a fraction of the needs of the Requirements Engineer and should be seen as one text among many. In particular the author views this book as complementary to Davis (1993) and Loucopoulos and Karakostas (1995).

NOTES

1   The term 'customer' refers to the person(s) who commission and/or purchase a product from a supplier; 'supplier' refers to the person(s) who develop the product and deliver it to the customer in exchange for payment.

APPENDIX A

Cost–Benefit Assessment of the Organisational Impact of a Technical System Proposal *

INTRODUCTION

Of the many tasks that have to be undertaken if information technology is to be successfully implemented one of the most crucial is the early assessment of the organisational impact of a proposed technical system. Typically the wider implications of technical system implementation are not appreciated until development work is well advanced and it may be difficult to change direction. The earlier this assessment can be made the more influence can be exerted upon technical system design and the methodology for its implementation. The problem, however, is how to make an assessment when the proposal is only an outline 'on the drawing board' and there is nothing concrete for users to evaluate.

The procedure presented here is designed to meet this need. The rationale and general structure is presented in Chapter 6[1] and the check-lists that follow provide the working documentation for the assessment. The circumstances under which this procedure is most relevant and has been most tested are as follows. As a result of a feasibility study an outline conceptual specification for a technical system has been proposed for a specific organisational setting. A group, perhaps comprising user management, technical staff and user representatives are charged with assessing this proposal for organisational and user acceptability.

The procedure takes a group of this kind through the following stages:

1. *Systems specification.* Stating the technical proposal in a form which facilitates the assessment of organisational impact.
2. *Organisational specification.* Outlining that part of the organisation

---

* Reproduced with permission from K. D. Eason (1988), *Information Technology and Organisational Change*, Taylor and Francis.

which will be affected by the system and describing the actual or
planned work roles that will be occupied by users of the system.

3. *User Cost–Benefit Assessment.* An assessment of the impact of the
   system upon the major work roles of potential users.
4. *Organisational Match Assessment.* An assessment of the overall
   impact upon the organisation.
5. *Socio–technical Design.* A series of check-lists to support the
   development of a strategy for the development of an acceptable
   socio–technical system.

In addition to its value in assessing specific technical proposals it has also
proved useful as a training aid, helping people appreciate the way in
which technical systems influence organisational issues and thereby
developing their ability to play a constructive role in technical developments within their own organisations. In this instance case studies need
to be developed with technical and organisational specifications similar
to the organisational contexts with which the trainees are familiar.

Stage 1: The Technical System Specification

The following information needs to be extracted from the outline conceptual proposal.

1. *Purpose and Overall Configuration*
   - What is the overall rationale for the system?
   - What scale is envisaged (how much of the organisation will be
     affected)?
   - What kind of system is envisaged (an integrated, mainframe-
     based system a net work, stand-alone micros, etc.)?
   - How does the system relate to existing technical systems, i.e., does
     it replace them, extend them, have to be compatible with them, etc?

2. *Planned Benefits*
   - What benefits are used to justify the planned expenditure?
   - resource reduction?
   - resource effectiveness?
   - individual enhancement?
   - organisational enhancement?
   - What priorities are placed upon the achievement of these benefits?

3. *System Functionality*
   - What, in global terms, are the main categories of service to be
     offered to users:
   - reports/enquiry facilities available?
   - communication facilities?

- text processing services?
- data processing and calculation services?
- control facilities, e.g. for monitoring and directing equipment, and processes? etc.
- What requirements, in global terms, will this place upon users:
- requirements for data input to the system?
- requirements for standardisation procedures across the organisation?
- requirements for timing, security, etc?

4. *Management Development, Management Control*
- From what location will the completed system be managed both on a routine and a developmental basis?
- From what location will the system be developed and what plans exist for the management of the project?
- What kind of development and implementation strategy is envisaged (e.g. prototypes and trials, phased implementation, 'big bang', etc.)?

It should be noted that very little information about the physical characteristics of the system is sought at this stage; indeed such information should not be available because decisions of this kind should follow a detailed examination of the concepts being advanced. This means that the assessment of impact can cover functionality and acceptability but not usability since this depends upon the specific form of system delivered. However, the analysis of users to be undertaken can lead to the statement of usability criteria[2] to be used as the system is developed.

Stage 2: Organisational Description

A statement of the characteristics of the organisational setting into which it is planned the technical system will be implemented.

1. *Organisational structure and work roles*
- What is the reporting structure in the part of the organisation to be affected by the system?
- What are the main categories of work role to be affected by the systems?
  - primary users?
  - secondary users (occasional or indirect users)?
  - tertiary users (those affected but without direct access)?
  - non-users but affected by implications, e.g. job displaced by the system?
- What changes are envisaged in the work roles or reporting structure by the time the system is implemented (either as a result of the system or because of other organisational changes)?

2. Overall allocation of relevant tasks to work roles

A top level task description of the way in which the activities of the organisation that the technical system is designed to support are currently handled in the organisation. It should show:

- How the overall tasks are subdivided and allocated to existing work roles (how are responsibilities allocated?)
- The existing contribution of technology to the tasks undertaken within each work role.
- The task interdependencies between the work roles. It may be that these can be shown as a time-related task-flow, as a complex process is worked through. However, care must be taken not to assume a too simplistic or well structured set of relationships. An analysis of the relationships between responsibilities may provide a more abstract and powerful expression of the interdependencies.

3. An Allocation of Functionality Table

The development of a table expressing the planned allocation of the functionality of the technical system to the work roles in the organisation. This brings together the system specification and the organisational specification and provides the basis for the cost–benefit assessment.

- Take each major role and list the system functionality which will be available to the role.
- Annotate the list with the allocation of responsibilities for data inputs and any other requirements which will be laid upon the work role.

System functionality	User work roles		

FIG. A.1 An allocation of functionality table.

Stage 3: User Cost–Benefit Assessment

The following check-lists can now be completed on behalf of each of the user work roles identified in the allocation of functionality table. The aim is to express the changes that the planned system would bring to each user role and to assess how the users would be likely to evaluate these changes.

In making the evaluation the underlying rationale is that users will be concerned with two aspects of the change:

Benefits: which may be in terms of the ability to perform tasks more effectively, desirable changes in the nature of the job, improved salaries, greater power and influence or enhanced career prospects, etc.

Costs: which may be financial but include the effort it takes to use a system, loss of job security, effort to learn and adapt, risk of failure, loss of job satisfaction and loss of privacy, etc.

The check-lists cover the major changes that research has shown often result from the introduction of information technology. The list is divided into five kinds of impact that range from the most direct to the most indirect. In practice, it is easier to predict the effects at the top of the list than at the bottom. The first issue is job security because when there are issues of this kind affecting a user group they will tend to dominate the entire evaluation process. Supposing a job remains to be done, the next section reviews the changes in information service that result from the functionality in the system which will be provided for the user group. The main categories of service need to be entered in this section. It is important to consider both the categories the access to and those that are provided because access by others to facilities debarred from the user group may have important consequences for them.

The third section reviews the major dimensions of a job that may be changed and this is followed by a section examining the organisational procedures that may change and influence the user. Finally, the indirect effects upon personnel policies which affect the user are reviewed.

In completing the questionnaire the significant changes that may affect the user are first listed. It is most unlikely that any specific user group will experience changes on all of these dimensions so that this is a question of trying to identify major areas of change. Where there are changes an assessment can be made of the likely evaluation of users. Separate columns are provided for benefits and costs because some changes have elements of both, for example, job changes can lead to the loss of valued skills but the development of new ones that will be valuable in the future.

It may be sufficient to give a qualitative expression to this evaluation, simply identifying all benefits as positive outcomes and all costs as negative outcomes. In practice, however, we have found that teams engaged in this exercise like to give quantitative expression to the evaluation. If this is done an overall score can be given for each user group and, across the user population, the winners and the losers can be clearly seen. We give below the scoring system that has evolved for this purpose but we would caution against too literal a use of the figures; the ability to predict changes and human responses to them is not such as to support fine discriminations between the totals that result.

To score the check-list put a total between +1 and +5 in the benefit column for each change that would be regarded as a benefit. A high score represents a change the user would be eagerly awaiting, a low score is a marginal benefit that would excite little interest. Similarly where a change may have a negative element to the user enter a score in the costs column of between −1 and −5 where −5 is a striking issue and −1 is a minor anxiety that may soon be forgotten. Averages can be calculated for each sub-section, treating costs and benefits separately. Dimensions where no changes are anticipated should be ignored. Totalling the scores for each subsection gives an overall score for the user group where the total possible is 25. Although the totals must be treated with caution a high benefit, low cost conclusion is indicative of a good response. A low benefit, high cost outcome will probably lead to user resistance and it may be useful to explore what form this might take (for example, non-use, partial, use, opposition to implementation, negotiation for protection or other benefits, etc). Quite often a high benefit, high costs outcome is obtained which suggests the reception of the system could be good if only some of the costs could be removed. In practice, it is common to find all of these outcomes for the same system because different user groups are affected in different ways.

The first check-list is an assessment of the probably outcomes of the planned system. It is quite likely that in completing the check-list respondents think of ways in which the outcomes for the users could be improved by changing the nature of the technical system, the way it is implemented or the social system of which the user is a part. A second check-list is provided for these desirable outcomes. It includes the same dimensions and can be scored in the same way. In this case, wherever a more desirable outcome is identified an 'If...' statement could be inserted to specify the conditions under which this desirable end would be obtained. A complete list of "If...' statements constitutes a useful specification of the requirements of the user groups, and can be used in replanning or detailing the system. There is of course no guarantee that the requirements of the different user groups will be compatible with one another.

USER COST–BENEFIT ANALYSIS

User group ....

Issues	Changes	Benefits	Costs
1. Job security			
2. Information facilities in system (system functionality) a) .... b) .... c) .... d) .... e)			
Average			
3. Job content a) Task variety b) Effort required c) New skills/old skills lost d) Work pace deadlines e) Workload f) Satisfaction			
Average			
4. Organisational procedures a) Discretion/autonomy b) Standardisation/formality c) Power and influence d) Privacy e) Communications f) Status			
Average			
5. Personnel policies a) Basic pay b) Other rewards c) Career prospects d) Industrial relations			
Average			
TOTALS			

FIG. A.2 Check-list 1: Problem outcome.

COST–BENEFIT ANALYSIS

User group ....

Issues	Changes	Benefits	Costs	Conditions
1. Job security				If...
2. Information facilities in system (system functionality) a) .... b) .... c) .... d) .... e)				If...
Average				
3. Job content a) Task variety b) Effort required c) New skills/old skills lost d) Work pace deadlines e) Workload f) Satisfaction				If...
Average				
4. Organisational procedures a) Discretion/autonomy b) Standardisation/formality c) Power and influence d) Privacy e) Communications f) Status				If...
Average				
5. Personnel policies a) Basic pay b) Other rewards c) Career prospects d) Industrial relations				If...
Average				
TOTALS				

FIG. A.3 Check-list 2: Desired outcome.

If the end product of this evaluation is a number of user groups who have very negative scores it is unlikely that the system can be effectively implemented except by management coercion. It may be, however, that the analysis has identified a number of more productive routes and an effective strategy can be to reformulate the technical plans and repeat the evaluation to check whether a better prognosis can be obtained. When the outcome is reasonably positive, i.e. all user groups show some positive outcomes even if there are considerable costs, stage 4 of the evaluation can be addressed.

Stage 4: The Assessment of Organisational Cost–Benefit

Check-list 3 provides a basis for making an assessment of the cost–benefit of the system from an organisational viewpoint. It is likely that a formal cost–benefit assessment will be made in terms of the value of tangible benefits set against the costs of system development and purchase. The aim here is to express the cost–benefit to the organisation across a wide range of organisational impacts.

The check-list has the same construction as the check-lists for user groups. It provides a number of dimensions for which the cost–benefit can be assessed by identifying the change and evaluating this for positive and negative implications.

The first section lists the major forms of planned benefits which were discussed in Chapter 2.[3] It would be expected that the main positive effects would arise from this section. However, the achievement of one type of benefit may well inhibit the achievement of other benefits which may be regarded as a negative outcome.

The second section lists the direct organisational implications of using information technology systems which are discussed in Chapter 8.[3] They range from issues of data security and reliability to the degree to which new systems are compatible with existing systems.

Broader organisational issues are considered in section three. A technical system may, for example, help or hinder the organisation's ability to respond flexibly to external demands, to adapt over time and to adopt values, control mechanisms, etc. of its choosing.

The final section summarises the likely reactions of the major user groups. The results of the earlier analysis can be summarised for all the user groups on a +5, –5 scale as a way of reviewing the ease of difficulty of implementing the system as planned.

Completing this check-list provides a basis for assessing whether the planned system as outline is likely to be beneficial and acceptable to the organisation as a whole. If it is, we can proceed to stage 5 which presents procedures for detailing and developing the system. If there are significant

ORGANISATIONAL COST–BENEFIT ANALYSIS

User group ....

Issues	Changes	Benefits	Costs
1. Planned benefits a) Resource reduction b) Resource optimisation c) Individual enhancement d) Organisational enhancement			
	Average		
2. System operation a) Reliability b) Security c) Compatibility d) Vulnerability to stoppages			
	Average		
3. Organisational match a) Control mechanisms b) Adaptability c) Flexibility d) Culture and values			
	Average		
5. User group responses a) .... b) .... c) .... d) .... e) .... f)			
	Average		
	TOTALS		

FIG. A.4 Check-list 3.

difficulties it may be appropriate to review the technical system outline, the organisational structure or the development plan to seek a better match with organisational needs. A new concept derived in this way may be evaluated by conducting stages 3 and 4 once again.

Stage 5: Developing a Strategy for Design and Change

The outline, conceptual plan which has been assessed as acceptable can now be detailed. It is likely that the analysis will have revealed a range of problems and a series of desirable outcomes. In developing the plans for the system and the way it is designed, these 'costs' and 'benefits' need to be carried forward to ensure the benefits are realised and the costs are eliminated or managed. The strategy that is developed should set this as a target for the organisation as a whole and for each of the major user groups.

The strategy has three components reflected in the three check-lists that follow.

The first step is to review the social system (and therefore the organisational change) that is appropriate to achieve the business goals of the enterprise. The second step is to construct an outline technical system which will support the social system. Finally, a project design process is required to construct and deliver the planned socio–technical system.

The check-list (check-list 4) for the social system is based upon Chapter 7[3] and part of Chapter 9.[3] It attempts to identify the overall enterprise goals and to structure the social system to serve these goals. This section may be about a major or a minimal change and should use the ideas generated in the analysis phase. It should be remembered that the assumptions of no change in the social system if a new technical system is implemented are not tenable. Therefore, even if no change is intended, this section should be reviewed because enforced changes may have to be managed. The second part of the check-list identifies the dimensions of the organisational changes that will take the enterprise from its current to its future state.

TABLE A.1. Check-list 4: Social system design.

Organisational Objectives

1. *Enterprise Goals:* state as operational aims in specific time-scales.
2. *Organisational Structure:* overall section/department/group structure to take responsibilities for enterprise goals.
3. *Job Design Philosophy:* policies with respect to the allocation of duties to individual employees.
4. *Control and Co-ordination:* policies to govern the way in which activities are co-ordinated in pursuit of goals.
5. *Values and Customs:* policies to govern the ways in which goals will be pursued.

Organisational Changes

For each of the following assess the degree of change and the appropriate mechanisms for making the transition.

1. Structural Changes in the Organisation.
2. Jobs to Phase Out.
3. New and Different Jobs to Introduce.
4. Skill Changes.
5. Implications for training, grading, team building, etc.
6. Industrial Relations Implications.

TABLE A.2. Check-list 5: Technical system specification.

Within the specification derived from check-list 4 develop a technical system specification which supports the social system which addresses the following topics for each user group.

1. *Facilities*
   - who needs what services from the system? (How is system functionality to be allocated?)
2. *Access*
   - who should have what degree of access to data bases?
   - how should access be controlled?
3. *Data Input*
   - by what means will data be entered, verified, etc?
   - who has the authority/responsibility for updating or amending which data?
4. *Interface Specification*
   - what special requirements will users have of the man–computer interface (ease of use, ease of learning, etc.) i.e. what will define usability for each user group?
5. *Customisation*
   - what requirements will there be for customisation of services to specific user groups?
6. *Adaptation*
   - what requirements will there be for adapting or evolving the system to meet developing user needs?
7. *Implementation Strategy*
   - what strategy will be used for implementation? e.g. prototypes, trials, phased introduction of facilities, etc?

TABLE A.3. Check-list 6: Project design process.

Considering the needs of (a) the management commissioning the system, (b) the technical staff and (c) each of the user groups, construct a process of designing the system by establishing the following:

1. *Steering Committee*
   - what is the customer–contractor relationship?
   - who makes policy decisions?
2. *Main Design Group*
   - should there be user representatives?
   - what is the role of user representatives?
   - should users be trained for systems design?
   - should it be full or part-time secondment?
3. *Working Parties*
   - should there be working parties on specific issues?
   - if so, which issues and what composition?
4. *Consultation, Training and Support*
   - who should consult and train each group of users?
   - what strategies should be used?
   - what point of need support strategy should be employed after implementation?
5. *Information Dissemination*
   - by what means will everybody be kept informed of design progress?
6. *Other Organisational Management Procedures*
   - What need is there to report to/consult existing management structures in the organisation?
   - Board of Directors
   - Information Technology Strategy Committee
   - Industrial Relations Bodies
   - etc.

Bear in mind that user participation has to be accomplished whilst the pressures to get the work done remain as normal.

Check-list 5 is a specification of those aspects of the technical system which most directly support the social system, i.e. support the role allocations between users. These issues are discussed in the user acceptability section of Chapter 8.[3] The check-list does not provide for a complete technical specification but concentrates upon the articulation of system services and requirements to match the articulation in the social system.

The final check-list, Check-list 6, seeks to establish the composition of the team who will develop the system and the procedures by which they will work. It is the design of the temporary vehicle which will manage the innovation. It is based upon Chapter 5[3] and concentrates upon the roles of users and technical staff in the development of the system.

The outcome of completing these check-lists should be a broadly-based specification of a socio–technical system which should serve significant enterprise goals and should be functional, usable and acceptable to the user groups within the organisation. Furthermore, the composition of the design process should assign responsibilities for this specification to users so that the system can be developed in detail without losing sight of the user and organisational variables that have been identified as significant.

NOTES

1   Chapter 6 refers to Eason's own book, Eason, 1988. Chapter 3, section 3.5 of the present book gave a summary of the rationale and the general structure of the technique.

2   See Macaulay, 1995, for an explanation of usability and usability criteria.

3   Of Eason, 1988.

APPENDIX B

Cooperative Requirements Capture
CRC Stage 1: A User Guide

INTRODUCTION

The aim of stage 1 of CRC is to enable stakeholders to develop a shared understanding of users and what they do now, and to project what users will be doing when the proposed system is introduced. Stage 1 also encourages team members to assess the feasibility of the proposed system and of the proposed change to the users situation, from the users' point of view.

CRC Stage 1 is normally conducted in a facilitated workshop setting. It would be attended by between 6 and 10 stakeholders and would be run by a facilitator. The workshop would typically last for two days.

Figure B.1 gives an outline of stage 1 of the CRC method showing the main steps. Step 1 is where the proposer of the new system presents the case for that system in terms of business opportunities and potential business benefits. The proposer gives an initial description of the proposed system and this is discussed with the other team members.

Steps 2, 3, 4 and 5 are similar. At each step, the team first of all produces a list by using brainstorming techniques. They then classify the contents of the list according to some criteria, then select one or two important items from the list, and describe those items using checklists of issues. The descriptions involve the team in thinking about the users' present jobs, the technological options and the proposed system.

Step 6 enables the team to look for interactions between the items and issues discussed in steps 2, 3, 4 and 5. Here, a number of short lists of user needs are generated.

Step 7 enables the team to reflect on the previous steps and to consolidate their findings into control sheets. These sheets are used by the team to help them assess the credibility of the information they currently have about users and to identify what further work they need to do in order to get a good understanding of users.

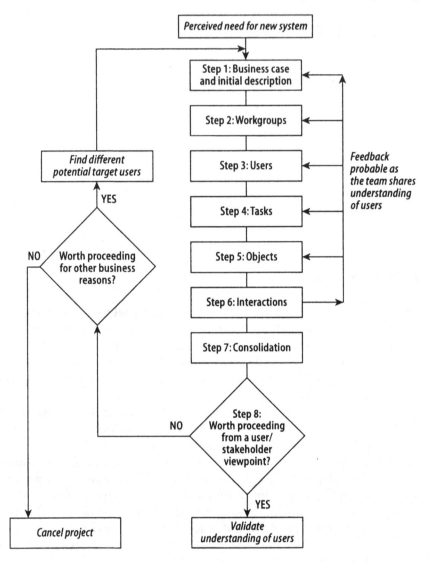

FIG. B.1 Stage 1 of the CRC method.

Step 8 is part of the consolidation process, in that, as a result of steps 2, 3, 4 and 5, the team will have discussed the proposed system and the proposed changes to the users' jobs and environment. The purpose of step 8 is to enable the team to identify the costs and benefits of the proposed system to the proposed users. In some cases it may be that little or no benefit can be identified and, thus, that it is not worth proceeding. The team may need to rethink the proposed project or else cancel it altogether.

If it is worth proceeding, then the team should undertake the further work identified at step 7: that is, they should validate their understanding of the users (after the meeting) and should complete the User Document.

STEP 1: BUSINESS CASE AND INITIAL DESCRIPTION

The business case consists of a statement, by one of the stakeholders, of the rationale for the system being proposed. This should include an initial description of the proposed system, an initial view on who the target users are and on the perceived benefits of the proposed change to the customers and users. In addition, the business case should identify the time perspective within which the proposed change will occur.

Thus, in the remainder of the discussion, *now* means 'at the time of the analysis' and *proposed* means *n* years from *now*. The analysis of change centres around the End Use Analysis form, in which the requirements team write down the current situation in the *now* column and the projected future situation in the *proposed* column. See Figs B.3, B.5 and B.6 for examples of End Use Analysis Forms.

STEP 2: WORKGROUPS

Each of the discussions concerning workgroups, users, tasks and objects includes: a brainstorming session, an evaluation session, a prioritisation session, and an analysis of change session. At the workgroup level, the team:

i.   identifies the workgroups associated with domain of interest by brainstorming suggestions for workgroups and writing them on to the whiteboard. Next, they discuss the list produced until agreement is reached that it represents the collective view; then the team

ii. classifies the workgroups according to whether they are likely to be primary, secondary or tertiary users of the proposed system (see Fig. B.2),

iii. Next, the team selects one primary workgroup and describes the workgroup as it is *now* in terms of organisational and social issues (see Fig. B.3).

Social issues are considered in order to help the designers to understand social aspects of the workgroup which may affect the acceptability of the proposed system to that group. They are considered in terms of workgroup structure (size of group, location, communication structure), workgroup dynamics (leadership styles and relationships within the group), and workgroup status and prestige, as perceived relative to other workgroups.

Work group title	Title 1	Title 2	Title 3
Relationship	☐	☐	☐
Generic Users ☐	☐	☐	☐
	☐	☐	☐
	☐	☐	☐
	☐	☐	☐
	☐	☐	☐
	☐	☐	☐

☐　Indicates the likely relationship of user or work group to the proposed system

1　Primary relationship: Likely to be frequent, hands-on, users

2　Secondary relationship: Likely to be occasional users or to use the system through an intermediary

3　Tertiaty relationship: Probably will not use the system, but may be affected by its use, or may influence its purchase

FIG. B.2 A workgroup table.

Organisational issues are considered in order to understand the workgroup in its organisational context, and to help anticipate any impact upon the organisation which may result from changes in working practice. Organisational issues are considered in terms of the workgroup's mission and objectives, its importance relative to other workgroups, the extent of its autonomy, its cohesion as a group, and the extent of its dependency on other workgroups.

iv. Next, for the same workgroup, the team considers job issues and attempts to describe how the workgroup will change in the *proposed* situation.

Consideration of job issues takes the form of a 'diary', a chart showing the time spent on particular activities within a suitable time period. It describes typical activities of the workgroup which are undertaken in a typical day, week or month (as appropriate). In some cases, a suitable

Proposed system _____

Workgroup _____

Organisational & social issues _____

Now	Proposed (e.g. 2 years on)
Mission/Objectives	
Autonomy	
Cohesion	
Dependency on other workgroups	
Structure and dynamics	
Prestige	

FIG. B.3 The workgroup: Organisational and social issues.

time period to consider might be the time taken to complete a typical project. The job issues would then include a list of typical activities which occur in the lifetime of a project.

STEP 3: USERS

At the user level, a similar procedure as that described for workgroups is followed. A list of generic users is agreed upon, and these are classified according to their relationship with the proposed system (see Fig. B.4).

First, the team selects one or more primary users and describes them according to three sets of issues:

i. The first set of issues is concerned with how the organisation views the generic user, both *now* and *proposed*. Consideration of organisational issues will determine the potential reactions of the organisation, and the possible impact upon the organisation that may result from introducing information technology into a particular user's job or occupation.

Generic Users	Relationship	Description sheets		
		Job	**Person**	**Organisation**

1: Primary user	YES: Means that a Description Sheet has been completed for that user
2: Secondary user	
3: Tertiary user	NO: Means it hasn't

FIG. B.4 List of generic users.

Organisational issues are considered in terms of the mission and objectives of the user, the importance of the user to other users within the organisation, the investment the organisation has made in the user, and the replaceability of the user by the organisation.

ii. The second set of issues is concerned with the personal attributes of the generic user, both *now* and *proposed*.

Person issues are examined to identify user characteristics that may affect the design and presentation of the proposed system, to understand the potential impact of the proposed system on the user's job or work, and to anticipate any adverse user reactions to the introduction of the proposed system into the workplace.

Person issues (see Fig. B.5) are considered in terms of the user's attitude and motivation towards work, information technology etc., the user's aspirations and ambitions, kinds and levels of skill and expertise possessed, and characteristics of the user's job: for example, whether it is lonely or sociable, dirty or clean, undertaken in hot or cold conditions, and so on.

iii. The third checklist, Fig. B.6, is a typical 'day in the life' of the generic user, both *now* and *proposed*. Job issues are considered in the same way as for the workgroup, in terms of a diary or a chart of time spent on various activities.

The 'day in the life' begins to identify where benefits should accrue from the use of the proposed system: for example, less time spent in meetings, faster corrections and changes to designs, less manual administration and record keeping. The team can thus begin to see which tasks are crucial to the success of the system in the eyes of the user.

STEP 4: TASKS

A task is defined as an action carried out by a generic user on an object in order to achieve a work goal. The procedure is as follows:

i. A list of tasks associated with a workgroup is first identified (see Fig. B.7). The allocation of function between human and computer is considered for each of the tasks.

ii. One or more primary tasks is selected and described using the check-lists. This will help the team to gain some understanding of the details of the task and the likely effect of change. Selected tasks are described using check-lists of organisational issues, timing issues and human issues, both *now* and *proposed*.

Proposed system _____

Generic user _____

Person issues _____

Now	Proposed (e.g. 2 years on)
Attitude	
Motivation	
Aspiration/Ambition	
Expertise	
Skill	
Job	

FIG. B.5 Person issues: generic user check-list.

Proposed system _____

Generic user _____

Job issues _____

Now	Proposed (e.g. 2 years on)

FIG. B.6 Job issues: A typical 'day in the life' of a generic user.

Task i.d.	Relationship	Description sheets		
		Organisation	Timing	Human

1:	Automated task: Likely to be totally automated, and the user will not take any action	YES:	Means that a Description Sheet has been completed for that user
2:	Shared task: Likely to be shared between the user and the system	NO:	Means it hasn't
3:	External task: Likely to be totally carried out by the user without system support		

FIG. B.7 A list of tasks, and the likely role of the system in supporting each task.

iii. A task is described in terms of organisational issues (see Fig. B.8): its importance from a political viewpoint within the organisation; its significance for security; the motivation for the user to carry out the task; the type of training required; and the other tasks supported by or supportive of the task.

Proposed system _____

Task _____

Organisational issues _____

Now	Proposed (e.g. 2 years on)
Importance	
Security	
Motivation	
Skill level	
Dependencies	

FIG. B.8 Task description sheet: Organisational issues.

iv. The task is also described in terms of timing issues (see Fig. B.9): that is, its frequency (i.e., how often the task is carried out by the user), the amount of time spent on it, the amount of preparation required to do it, and the degree of fragmentation of the task.

Proposed system _____

Task _____

Timing issues _____

Now	Proposed (e.g. 2 years on)
Time spent on:	
Task/Frequency	
Preparation	
Fragmentation	

FIG. B.9 Task description sheet: Timing issues.

v. In terms of human issues (see Fig. B.10), the team should consider the level of discretion the user has in deciding whether or not to carry out the task, the amount of stress involved in the task and the performance criteria used to measure successful completion of the task.

Proposed system _____

Task _____

Human issues _____

Now	Proposed (e.g. 2 years on)
Task discretion	
Task stress	
Performance criteria	

FIG. B.10 Task description sheet: Human issues.

At the task level the team gains a greater understanding of the users and the tasks they carry out, and can begin to identify which tasks will be of importance to the proposed system. In addition, the team begins to consider the role of the proposed system in supporting the tasks, and to identify what the learning needs of the user are with respect to learning the new system-supported tasks.

STEP 5: OBJECTS

At the object level the team is asked to carry out the following procedure:

i. Identify a list of objects associated with the users' environment. These objects will normally be associated in some way with users and workgroups. They could be real-world objects, knowledge about real-world objects, procedures remembered by users or other, more abstract, objects.

Once a list has been produced it is then reconsidered and revised by the team. This entails clarifying the meaning of object names, looking for similar objects with different names or two different objects with similar names. In addition, it may be possible to aggregate some objects with others: for example, some 'objects' may actually be attributes of other objects.

ii. Once an agreed list is produced the objects are then classified according to their degree of interest to the proposed system (see Fig. B.11).
iii.Selected objects are then described in further detail in terms of their *now* and *proposed* characteristics (see Fig. B.12).

At the object level the team begins to identify which objects in the user environment are likely to become objects which need to be supported by the system.

STEP 6: INTERACTIONS

Step 6 enables the team to look for interactions between the items and issues discussed in steps 2, 3, 4 and 5. Here, a number of short lists of user needs are generated. In particular, combinations of user, object and task are examined in order to assess needs or requirements associated with the proposed system. The team is encouraged to make statements of the form 'There is a need for…' – for example, 'There is a need for version control' – as opposed to 'The *xyz* system of version control will be implemented'. The purpose of this is that the team should be trying to identify user and customer needs, rather than deciding on the solution (see Fig. B.13).

Object i.d	Relationship	Description sheet

1: Hidden object:
   Likely to be totally automated, and not visible
   at the user interface

2: Visible object:
   Likely to be supported by the system, but will
   be visible to the user at the user interface

3: External object:
   Not likely to be supported by the system, likely
   to remain external to the system, but could
   still be of interest to the user

FIG. B.11 A list of objects, and the likely role of the system in supporting each object.

Proposed system _____

Object _____

Object description _____

Now	Proposed (e.g. 2 years on)
Description	
Access to the object	
Management	
Representation	
Quality	

FIG. B.12 Object description sheet.

User i.d.	
Object i.d.	
Task i.d.	
Interaction characteristics: A need for...	

FIG. B.13 User, object and task interactions.

STEP 7: CONSOLIDATION

Step 7 enables the team to reflect on the previous steps and to consolidate their findings into control charts. These charts are used by the team to help them assess the credibility of the information they currently have about users and to identify what further work they need to do in order to get a good understanding of users. In particular, the consolidation session includes a review of each of the workgroups, users, objects and tasks, in which the team is asked to make an honest assessment of the accuracy of their collective knowledge of the users (see the Control Sheets in Figs B.14, B.15 and B.16).

The team is then encouraged in step 7 to identify follow-up investigations that are needed in order to ensure that the future stages of requirements capture and analysis, and the system design, are based on a sound understanding of the users. The User Document is initially a collection of the proformas completed at the workshop, but should be expanded after the workshop.

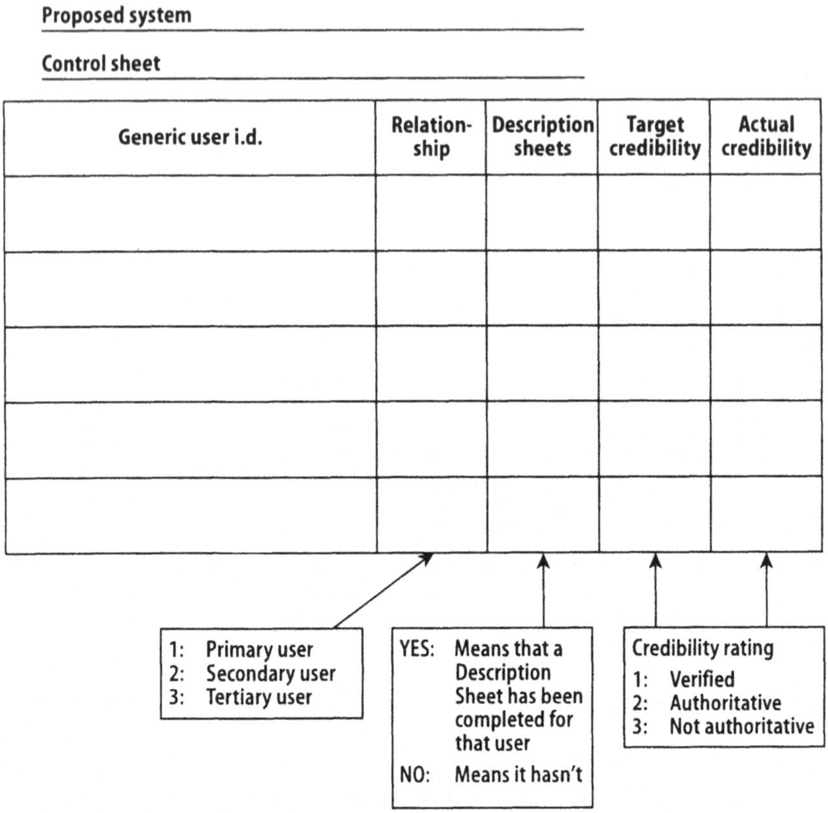

Proposed system _____

Control sheet _____

Generic user i.d.	Relation-ship	Description sheets	Target credibility	Actual credibility

1:   Primary user
2:   Secondary user
3:   Tertiary user

YES: Means that a
        Description
        Sheet has been
        completed for
        that user
NO: Means it hasn't

Credibility rating
1:   Verified
2:   Authoritative
3:   Not authoritative

FIG. B.14 Control sheet for generic users.

Proposed system

Control sheet

Object i.d.	Relation-ship	Description sheets	Target credibility	Actual credibility

1: Hidden object	YES: Means that a	Credibility rating
2: Visible object	Description	1: Verified
3: External object	Sheet has been	2: Authoritative
	completed for	3: Not authoritative
	that object	
	NO: Means it hasn't	

FIG. B.15 Control sheet for objects.

Proposed system _____

Control sheet _____

Task i.d.	Relation-ship	Description sheets	Target credibility	Actual credibility

1: Automated task	YES: Means that a	Credibility rating
2: Shared task	Description	1: Verified
3: External task	Sheet has been	2: Authoritative
	completed for	3: Not authoritative
	that task	
	NO: Means it hasn't	

FIG. B.16 Control sheet for tasks.

STEP 8: WORTH PROCEEDING?

Step 8 is part of the consolidation process, in that, as a result of steps 2, 3, 4 and 5, the team will have discussed the proposed system and the proposed changes to the users' jobs and environment. The purpose of step 8 is to enable the team to identify the costs and benefits of the proposed system to the proposed users. In some cases it may be that little or no benefit can be identified, and thus that it is not worth proceeding. The team may need to rethink the proposed project or else cancel it altogether. Or, as is sometimes the case, the team will proceed with the project but seek to identify different target users.

If is worth proceeding, the User Document provides the requirements team with an agreed set of descriptions of the target users and with an initial description of the requirements associated with the proposed system. The next stage in the process is to confirm that the descriptions of users are valid and to modify the User Document accordingly.

The starting point for the next workshop is a validated User Document. The team can then proceed to identify the scope of the proposed system by following the steps within Stage 2 of CRC.

The structure of the User Document is shown in Table B.1.

TABLE B.1 User Document: list of contents.

---

1. Management Summary
(including the business case and a brief description of the proposed system)

2. Organisation/Workgroups
    2.1 Workgroup Control Sheets
    2.2 Organisation Chart
    2.3 Workgroup Table
    2.4 Workgroup Description Checklists

3. Generic Users
    3.1 Generic Users' Control Sheets
    3.2 Generic Users' Description Checklists

4. Tasks
    4.1 Task Control Sheets
    4.2 Task Hierarchy
    4.3 Task Description Checklists

5. Objects
    5.1 Objects Control Sheets
    5.2 Object Structures
    5.3 Object Description Checklists

6. Interactions
    6.1 User/Task/Object Interactions
    6.2 Initial List of Requirements and Attributes

7. Consolidation
    7.1 Statement of Credibility
    7.2 Further Investigations Needed

8. Worth Proceeding?
    8.1 User/Stakeholder Perspective
    8.2 Business Perspective
    8.3 Plan of Action

9. Conclusion

APPENDIX C

Cooperative Evaluation
a Run-time Guide*

---

1. INTRODUCTION

1.1 What is Cooperative Evaluation?

Cooperative Evaluation is a procedure for obtaining data about problems experienced when working with a software product, so that changes can be made to improve the product.

1.2 Who Uses Cooperative Evaluation?

Cooperative Evaluation can be used by designers without specialised knowledge of human factors research.

1.3 When to Use Cooperative Evaluation

Cooperative Evaluation is most useful for early feedback about redesign in a rapid iterative cycle. The aim is not to provide an exhaustive list of all the problems that could possibly be identified. Rather, it is to help you identify, with the minimum of effort, the most important problems to consider. Cooperative Evaluation can be used with:

- an existing product that is to be improved or extended;
- an early partial prototype or simulation;
- a full working prototype.

---

* Reproduced with permission from A. Monk, P. Wright, J. Haber and L. Davenport (1993), *Improving Your Human–Computer Interface: A practical technique*, Prentice-Hall International.

1.4 About this Guide

This guide is intended as a 'stand alone' reference guide to Cooperative Evaluation. It is intended to help you prepare and run a Cooperative Evaluation session. It is not a detailed description of the technique, this can be found in Chapter 2.[1] Rather it is a series of questions, summaries, reminders and checklists for each stage of preparing and running a session. We suggest that the best way to use this guide is to photocopy it and use the checklists when appropriate as you progress through the procedure. The guide is divided into the three steps discussed in Chapter 2,[2] These reflect the main activities of preparing and running an evaluation session.

- Recruit users
- Prepare tasks
- Interact and record

2. RECRUIT USERS

2.1 Define the Target User Population

Before you can say whether someone is typical or atypical of the eventual users of the product you have to define who those eventual users will be. If the product is to be used by a specified department or group of individuals then the existing employees define this population. Simply write the name of the group in the box below. The development of more generic products will be preceded by market research which will specify the user group. Even if the population is not defined for you, make an explicit decision to aim at some target user population.

```
Who are the eventual
end users of this product?
```

2.2 Recruit Users who are as Similar to the Target User Population as is Practical

We suggest you work with between 1 and 5 users per iteration early in the design process. Precisely how many will depend on practical requirements such as: how much time you have, how long it will take to run each session (see Prepare Tasks below) and how much time your users have.

How many users will you work with?

Where you get your users from will depend on who you have specified as the target user population in Section 2.1. The figure below suggests some possibilities.

Target population	Suggested source of users
A specified company department or group of people	Ask the company, department or group
The general public	Advertise in newspapers
People with specific experience or skill	Use recruitment or secretarial agency
Don't know	Make up a 'user profile' and recruit people who fit the description

2.3 Things to Watch Out For During Recruitment

1. If you are NOT recruiting directly from an identified group, have you checked that the people you recruit have the same characteristics as your target users? Important considerations are:

   - Their knowledge of the task domain
   - Their experience of computers
   - Their skill at using keyboards and other input devices
   - Their level of education and how they will approach situations that require problem solving

2. Do you need to make arrangements to pay users? Do you need to seek permission for them to take time out of their normal work?

Where will you get your users from?
Whose permission will be required?
What administrative arrangements will need to be made?

3. PREPARE TASKS

Selecting the right tasks is crucial for the success of Cooperative Evaluation. They must be do-able by the users, they must be representative of real tasks the users do and they must explore the prototype thoroughly.

The aim of this step is to prepare a task sheet. This will contain a list of tasks that all your users will attempt to work through with your prototype. The task sheets are given to the user at the beginning of the Cooperative Evaluation session.

QUESTIONS TO ASK YOURSELF	
Questions	Answers and comments
Have you made sure that the tasks you have planned can actually be done using your prototype? Are they going to focus the user on the parts of the prototype you are interested in? How much time have you allowed for each user in total? How long do you estimate it will take each user just to complete the tasks? Is the total time you have allowed at least 50% greater than the time to complete the tasks? Have you written down the tasks in a way that can be understood by a novice user?	

3.1 Things to Watch Out For when Preparing Tasks

1. Are the tasks specific? 'Do your normal work' is not a specific task. 'Draw a house with a door, four windows and a chimney' is.
2. How do you know that the tasks you have chosen are representative of the work the product is designed to support? You may have talked to users about their job, you may have spoken to market research or other specialists in the company.
3. What are you going to do if the user cannot finish a task or if a user finishes too quickly? You may want to decide on a maximum time for

each task. You may want some extra tasks up your sleeve that are easier or that can be given to people who finish more quickly than you expected.

4. Important functions should be examined twice, once at the beginning and once at the end of the session.

Notes

4. INTERACT AND RECORD

This session is divided into four parts, telling what you need to do:

- Before the users arrive
- When the users arrive (before starting the tasks)
- While the users are using the system
- Debriefing the users

4.1 Before the Users Arrive

Everything needs to be in place, tested and fully operational. Use the following two checklists. The first lists all the things you need. The second lists everything you should have done.

Have you got the following?	Answers and comments
Your prototype ready to use in a reasonably quiet environment?	
A sheet containing the user's tasks?	
Some means of recording what the user says, ideally a clip-on microphone plugged into the video recorder? (see below)	
Some means of recording what the user does (i.e., a video recorder or a system logger)?	
A notebook or proforma sheet on which to make notes?	
A list of questions to ask during the debriefing?	

Have you done the following?	Answers and comments
Planned what you need to say when the user arrives? (see next section) Worked through the task sheet yourself so you know what to expect? Checked that the recording apparatus is working correctly?	

Notes

4.2 When the Users Arrive

The whole session should be conducted in an informal manner and you and the users should discuss the system openly. They should be encouraged to think of themselves as coevaluators not as experimental subjects. They should be told that you are interested in the way the system *misleads them* rather than in the *mistakes they make*. They should be told you are interested in the things that *the system makes it hard for them to do* rather than the things that *they are unable to do*. This emphasises that it is the system that is being evaluated not the user. This will help the question–answer dialogue to flow easily.

When the users arrive and before they start work on the tasks there are five things you need to do.

1. Put the users at their ease.
2. Start recording the session. Do this early on in case you forget later.
3. Introduce yourself. Explain who you are and the purpose of the session in general terms. Emphasise that you are testing the prototype system not the user. Explain the philosophy of Cooperative Evaluation.
4. Explain Cooperative Evaluation. Explain that what the user says will be taped and that everything is confidential.
5. Introduce the task sheet. Explain that the tasks are not a test, just a way of introducing the user to the various parts of this new system.

We think this part of the session is best done informally so we DO NOT recommend that you have a written set of instructions. But we included some written instructions below to give you an idea of what you need to say.

SAMPLE INSTRUCTIONS FOR COOPERATIVE EVALUATION

Thank you for agreeing to help with this study. Today we are going to evaluate the usability of a particular computer system called REF.

REF stores large numbers of references to academic works rather like the catalogues in a library. It can be used to search for things written on a particular subject, or by a particular author and so on.

The aim of the study is to find out how easy REF is to use by people like yourself. We want you to use it to help us find out what problems REF poses and how it could be improved.

We will give you some standard tasks to do using REF. The aim of this is to allow us to get some information about how REF supports this activity. We are particularly interested in situations in which REF encourages you to make errors in selecting commands and misleads you about what it will do. We are also interested in extra commands that would make the system easier to use.

To get this information we shall use a question-and-answer technique. This involves three things.

1. We want you to think-out-loud as you do each task, telling us how you are trying to solve each task, which commands you think might be appropriate and why, and what you think the machine has done in response to your commands and why. Think of this as you give us a running commentary on what you are doing and thinking.

2. Whenever you find yourself in a situation where you are unsure about what to do or what effect commands might have, ask us for advice. If you ask us what you need to know we will suggest things for you to try but if you get really stuck we'll explain exactly what you have to do.

3. In addition we will ask you questions about what you are trying to do and what effect you expect the commands you type will have. This is simply to find out what problems there are with the system. During our conversations, we want you to voice any thoughts you have about parts of the system which you feel are difficult to use or poorly designed.

While you are doing this we will be noting down the problems you mention but in case we miss any we are going to audio tape our conversation. This recording will be anonymous and treated in confidence.

Remember it is not you we are testing, it is REF. We are interested in what you think so do not treat this as an examination. Treat it as a structured discussion about REF. Please feel free to say whatever you think about the system and the tasks you are given to solve.

4.3 While the Users are Using the System

The two main things to remember to do when the users are actually carrying out the task are:

Keep them talking!
Make sure you know what's going on!

---

1. Encourage the users to think aloud while using the system. This is best done by asking them to give you a running commentary of what they are doing and what's going on.
2. Ensure that there is a relatively continuous dialogue by asking appropriate questions whenever possible. Section 4.4 is a list of useful questions. These can be photocopied and used as a 'crib sheet' during the session.
3. Note each occurrence of unexpected behaviour and each comment on the usability of the system. You will not have time to make detailed notes. What you need to do is make a note of where on the tape the behaviour or comment occurred and a brief (possibly one word) description of what happened. A proforma for this is included at the end of this section. Do not let note taking interfere with the primary task of creating a dialogue with the user. Stop note taking rather than let this happen.

    Unexpected Behaviour is where the users do something the designer did not intend them to do. For example, the user might type in an inappropriate sequence of commands or data.

    Comments are subjective comments or evaluations of the interface. These can be both positive and negative ('It's nice the way you can do that without having to type the whole thing again', 'That seemed to take a lot of effort', 'I don't like having to do that twice' and so on).

4.4 Some Useful Questions to Ask During the Evaluation

- How do we do that?
- What do you want to do?
- What will happen if...?
- What has the system done now?
- What is the system trying to tell you with this message?
- Why has the system done that?
- What were you expecting to happen then?
- What are you doing now?

4.5 Debriefing the Users

When the user has finished the set tasks you should spend some time talking about the session. Keep the tape recorder on during this time. Some very interesting comments emerge out of this part of the session. As well as discussing what each of you think are important usability problems you can also get some feedback about the Cooperative Evaluation session itself. Some useful questions to ask the user are included in Section 4.6. These can be photocopied and used as a prompt. These question are, however, very general; you will probably want to ask some fairly specific questions of your own about specific aspects of the prototype such as menu names, default values and so on.

If you are testing many users or you want more formal feedback, it may be worth considering drawing up a simple questionnaire for the users to fill in. Be careful not to make it too long or complicated. A good example of such a questionnaire is included in Chapter 3 as Figure 3.3.[3]

It is sometimes possible to see users a second time either individually or as a group. If this is possible it is very useful. It allows you the opportunity of clarifying your interpretation of important usability problems and also discussing possible design changes. It also serves a useful customer relations function bringing users and designers together for a round-table discussion. This is discussed in more detail in Chapter 3.[3]

4.6 Some Useful Questions to Ask During Debriefing

About the prototype:

- What do you think was the best thing about the prototype?
- What do you think was the worst thing about the prototype?
- What do you think most needs changing?
- How easy did you find the tasks?
- Specific questions about the prototype...

About Cooperative Evaluation:

- Did you find the recording equipment intrusive?
- Were the tasks similar to things you currently do?
- How realistic did you find the prototype?

Use the sample proforma sheet below for making notes

No.	Tape count	Unexpected behaviour or comment
1		
2		
3		
4		
5		
6		
7		
8		
9		
10		
11		
12		

NOTES

1. Chapter 1 of Monk *et al.*, 1993.

2. And as explained in Chapter 5, section 5.5

3. Chapter 3 of Monk *et al.*, 1993

REFERENCES

August, J. H. (1991). *Joint Application Design The Group Session Approach to System Design*. Englewood Cliffs, NJ: Yourdon Press.

Avison, D. E. and Wood-Harper, A. T. (1991). 'Information systems development research: an exploration of ideas in practice.' *The Computer Journal* 34(2): 98–112.

Bantleman, J. P. (1985). A feature analysis of L.S.D.M., in *Structured methods: State of art report*, pp 5–25. Pergamon.

Bell, F. and Oates, B. (1994). 'A Framework for Method Integration'. *Second BCS Conference on Information Systems Methodologies, Edinburgh*. British Computer Society.

Berrisford, R. A. (1993). *Object-oriented SSADM*. Prentice-Hall.

Betts, M. (1989). 'QFD integrated with Software Engineering'. *The second symposium on Quality Function Deployment*, GOAL/QPC, (op. cit.).

Beyer, H. R. and Holtzblatt, K. (1995). 'Apprenticing with the customer', *Communications of the ACM* 38(5): 45–53.

Bhabuta, L. (1989). 'Balancing System and Organisational Needs: User Involvement in Requirements Analysis', in K. Knight (ed.), *Participation in Systems Development*, pp. 134–51. Kogan Page.

Bjorn-Anderson N., Eason K. and Robey D. (1986). *Managing Computer Impact: An International Study of Management and Organisations*. Ablex Publishing Corporation.

Bødker, S. and Grønbaek, K. (1991). 'Design in Action: From Prototyping by Demonstration to Cooperative Prototyping', in J. Greenbaum and M. Kyng (eds), *Design at Work: Cooperative Design of Computer Systems*. Lawrence Erlbaum.

Booch, G. (1991). *Object-Oriented Design with Applications*. Redwood City, CA: Benjamin/Cummings.

Bossert, J. L. (1991). *Quality Function Deployment: A Practitioner's Approach*. Milwaukee, Wis: ASQC Quality Press.

Brandt, I. (1983). 'A comparative study of information systems design methodologies', in T. W. Olle *et al.* (eds), *Information system design methodologies: a feature analysis*. North-Holland.

Bravo, E. (1993). 'The hazards of leaving out users', in D. Schuler and A. Namioka (eds), *Participatory Design: Principles and Practices*, pp. 3–13. Hillsdale, NJ: Lawrence Erlbaum.

British Standards Institution (1986). *British Standard Guide to Specifying user requirements for a computer-based system* (BS 6710).

Brodie, M. L. *et al.* (1983). 'On a framework for information system design methodologies', in T. W. Olle *et al.* (eds), *Information system design methodologies: a feature analysis*. North-Holland.

Bubenko, J. A. (1995). 'Challenges in Requirements Engineering'. *Second IEEE International Symposium on Requirements Engineering, York*, pp. 160–4. Los Alamitos, Cal: IEEE Computer Society Press.

Bustard, D. W. and Lundy, P. J. (1995). 'Enhancing Soft Systems Analysis with Formal Modelling'. *Second IEEE International Symposium on Requirements Engineering, York*, pp. 164–72. Los Alamitos, Cal: IEEE Computer Society Press.

Carmel, E. *et al.* (1993). 'PD and Joint Application Design: A Transatlantic Comparison'. *Communications of the ACM* 36(4): 40–8.

Catterall, B. J. (1990). 'The HUFIT functionality matrix', in D. Diaper, D. Gilmore, G.

Cockton and B. Shackel (eds), *INTERACT'90*, pp. 377–81. Cambridge: IFIP.

Chatzoglou, P. and Macaulay, L. A. (1995). 'Requirements Capture and Analysis: the Project Manager's Dilemma'. *International Journal of Computer Applications in Technology* 8(3/4).

Checkland, P. (1995).'Model Validation in Soft Systems Practice'. *Systems research* 12(1): 46–54.

Checkland, P. B. (1981). *Systems Thinking, Systems Practice*. Chichester: John Wiley and Sons.

Checkland, P. B. and Scholes, J. (1990). *Soft Systems Methodology in Action*. Chichester: John Wiley and Sons.

Coad, P. and Yourdon, E. (1991). *Object Oriented Analysis*. Yourdon Press.

Cohen, W. (1988). 'Quality Function Deployment: An Application Perspective from Digital Equipment Corporation'. *National Productivity Review*, Summer 1988: 197–208.

Crawford, A. (1994). *Advancing Business Concepts in a JAD Workshop Setting*. Englewood Cliffs, NJ: Prentice-Hall.

Crosby, P. (1985). *Quality Improvement through Defect Prevention*. Philip Crosby Associates, Inc.

Curtis, B., Krasner, H. and Iscoe, N. (1988). 'A Field Study of the Software Design Process for Large Systems'. *Communications of the ACM*, 31(11), November 1988.

Daetz, D. (1989). 'QFD: a method for guaranteeing communication of the customer voice through the whole product development cycle', in *IEEE International Conference on Communications, Boston, IEEE*, pp. 1329–33.

Dale, A. G. (1984). 'The ISAC approach to specification of information systems', in T. W. Olle *et al.* (eds), *Information system design methodologies: a comparative review*. North-Holland.

Darzentas, T. and Spyrou, T. (1993). 'Information systems for primary health care: the case of the Aegean islands'. *European Journal of Information Systems* 2(2): 117–27

Davis, A. M. (1993). *Software Requirements: Objects, functions and States*. Englewood Cliffs, NJ: Prentice Hall International.

de Bachtin, O. (1985). 'It is what it's used for – job perception and system evaluation', in B. Shackel (ed.), *Human–computer interaction – INTERACT '84*. North-Holland.

DeMarco, T. (1979). 'Structured Analysis and System Specification'. Englewood Cliffs, NJ: Prentice-Hall.

Dobbin, T. J. and Bustard, D. W. (1994). 'Combining Soft Systems Methodology and Object Oriented Analysis: The search for a good fit', in Lissoni, C. *et al.* (eds), *Proceedings of the second conference on Information Systems Methodologies*, pp. 69–80. BCS ISM Specialist Group.

DOORS (1995). Quality Systems and Software, DOORS: Requirements Management Tool.

Douglas, T. (1970): *A Decade of Small Group Theory, 1960–1970*. London: Bookstall Publications.

Downs, E. *et al.* (1992). *Structured Systems Analysis and Design Method: Application and Context*. Prentice-Hall.

Draper, S. and Oatley, K. (1991). 'Practical Methods for Measuring the Performance of Human–Computer Interfaces'. *JCI Summerschool notes, 1991*.

Eason, K. (1992). *The development of a user-centred design process: a case study in multi-disciplinary research*. (Inaugural Lecture.) Loughborough University of Technology.

Eason, K. D. (1982). 'The Process of Introducing Information Technology'. *Behaviour and Information Technology*, 1(2), April–June 1982.

Eason, K. D. (1988). *Information Technology and Organisational Change*. Taylor and Francis.

Eason, K. D. (1989). 'Tools for Participation: How Managers and Users Can Influence Design', in K. Knight (ed.), *Participation in Systems Development*, pp. 94–105.

Kogan Page.

Ehn, P. and Kyng, M. (1991). 'Cardboard Computers: Mocking-it-up of Hands-on-the-future', in J. Greenbaum and M. Kyng (eds), *Design at Work: Cooperative Design of Computer Systems*, pp. 169–97. Hillsdale, NJ: Lawrence Erlbaum.

Ehn, P., Mollervd, B. and Sjogren, D. (1990). 'Playing in reality: A paradigm case'. *Scandinavian J. Inf. Syst.* 2: 101–20.

El Emam, K. and Madhavji, N. H. (1995). 'Measuring the Success of Requirements Engineering Processes'. *Second IEEE International Symposium on Requirements Engineering, York*, pp. 204–14. Los Alamitos, Cal: IEEE Computer Society Press.

Emery, F. E. and Trist, E. L. (1969). 'Socio-Technical Systems', in F. E. Emery (ed.), *Systems Thinking*. London: Penguin.

Fields, R. E., Wright, P. C. and Harrison, M. D. (1995). 'A Task Centred Approach to Analysing Human Error Tolerance Requirements'. *Second IEEE International Symposium on Requirements Engineering, York*, pp. 18–27. Los Alamitos, Cal: IEEE Computer Society Press.

Floyd, C., Mehl, W., Reisen, F., Schmidt, G. and Wolf, G. (1989). 'Out of Scandinavia: Alternative approaches to software design and system development'. *Human–Computer Interaction* 4: 235–50.

Flynn D. J., Layzell, P. J. and Loucopoulos, P. (1986). 'Assisting the Analyst – The aims and objectives of the Analyst Assist project'. *BCS Conference on Software Engineering, Southampton*.

Flynn, D. J. (1992). *Information Systems Requirements: Determination and Analysis*. McGraw-Hill Europe.

Gammack, J. G. and Young, R. M. (1985). 'Psychological techniques for eliciting expert knowledge. Research and Development', in M. A. Bramer (ed.), *Expert Systems*, pp. 105–12. Cambridge: Cambridge University Press.

Gasson, S. (1995). 'User involvement in decision-making in information systems development'. *Proceedings of IRIS 18, Gjern, Denmark, Gothenburg Studies in Informatics, Report 7*, pp. 201–17.

Gibson, C. F. and Jackson, B. B. (1987). *The Information Imperative: Managing the Impact of Information Technology on Business and People*. Lexington, Mass: Lexington Books D.C. Heath.

Glasson, B. C. (1984). 'Guidelines for user participation in the system development process', in B. Shackel (ed.), *Human–computer interaction – INTERACT '84*, North-Holland.

GOAL/QPC (1989). 'Quality Function Deployment: A Process for Continuous Improvement'. *The second symposium on Quality Function Deployment, GOAL/QPC, 13 Branch Street, Methuen, MA 01844*.

Gould, J. D. and Lewis, C. (1985). 'Designing for usability: Key principles and what designers think'. *Communications of the ACM*, 28: 300–11

Greenbaum, J. and Kyng, M. (eds) (1991). *Design at Work: Cooperative Design of Computer Systems*. Hillsdale, NJ: Lawrence Erlbaum.

Hart, A. (1985). 'Knowledge elicitation: issues and methods'. *Computer-Aided Design* 17(9): 455–62.

Hughes, J., O'Brien, J., Rodden, T., Rouncefield, M. and Sommerville, I. (1995). 'Presenting Ethnography in the Requirements Process'. *Second IEEE International Symposium on Requirements Engineering, York*. Los Alamitos, Cal: IEEE Computer Society Press.

Hutt, A. T. F. (1994). *Object Analysis and Design: Description of Methods*. John Wiley and Sons.

IEEE (1984). *IEEE Guide to Software Requirements Specifications, IEEE Std 830-1984*. IEEE Inc., 345 East 47th St, New York, NY 10017,USA.

IEEE (1990). Dorfman, M. and Thayer, R. H. (1990). *Standard, guidelines, and examples on system and software Requirements Engineering*. Los Alamitos, Cal: IEEE Computer Society Press.

Ince, D. C. *et al.* (1993). *Introduction to Software Project Management and Quality*

Assurance. McGraw-Hill.

Jarke, M. and Pohl, K. (1994). 'Requirements Engineering in 2001: (virtually) managing a changing reality'. *Software Engineering Journal*, Nov.

Jirotka, M. and Goguen, J. (1994). *Requirements Engineering: Social and Technical Issues*. Academic Press.

Johnson, P. (1992). *Human Computer Interaction Psychology, Task Analysis and Software Engineering*. London: McGraw Hill.

Jungk, R. and Mullert, N. (1987). *Future Workshops: How to create desirable futures*. London: Institute for Social Inventions.

Kano, N., Seraku, N., Takahashi, F. and Tsuji, S. (1984). 'Attractive and must-be quality'. (Japanese) *Hinshitsu* 14(2): 39–48.

Kawalek, P. (1983). 'BPR – A Revolutionary Manifesto?'. *Computer Bulletin*, the Newsletter of the British Computer Society, December 1993.

Kawalek, P. and Wastell, D. (1994). 'The Development of a Process Modelling Method'. *Second BCS Conference on Information Systems Methodologies, Edinburgh*, pp. 295–307. British Computer Society.

Keil, M. C. E. (1995). 'Customer–Developer Links in Software Development'. *Communications of the ACM* 38(4): 33–44.

Kensing, F. (1987). 'Generation of visions in system development', in P. Docherty *et al.* (eds), *System design for human productivity – participation and beyond*, pp. 285–301. Amsterdam: North Holland.

Kensing, F. and Madsen, K. H. (1991). 'Generating Visions, Future Workshops and Metaphorical Design', in J. Greenbaum and M. Kyng (eds), *Design at Work*. Hillsdale, NJ: Lawrence Erlbaum.

Kensing, F. and Munk-Madsen, A. (1993). 'PD: Structure in the Toolbox', *Communications of the ACM*, 36(4).

Kent, S. J. H, Maibaum, T. S. E and Quirk, W. J. (1993). 'Formally specifying temporal constraints and error recovery'. *Proceedings of the First IEEE International Symposium on Requirements Engineering*, pp. 208–16. Los Alamitos, Cal: IEEE Computer Society Press.

King, B. (1989). 'Better designs in half the time: Implementing QFD in America'. Methuen, Mass.: GOAL/QPC. (op. cit.)

Konda, S. *et al.* (1992). 'Shared memory in design: A Unifying theme for research and practice'. *Research in Engineering Design* 4(1): 23–42.

Kyng, M. (1991). 'Designing for Cooperation: Cooperating in Design'. *Communications of the ACM* 34(12): 65–73.

Lano, K. and Haughton, H. (1992). *The Z++ Manual*. Lloyd's Register.

Lano, K. and Haughton, H. (eds) (1993). *Object-oriented Specification Case Studies*. Prentice-Hall.

Lecuyer, D. (1989). 'Lessons learned from a QFD on the space transportation engine'. *The second symposium on Quality Function Deployment*, GOAL/QPC (op. cit).

Loucopoulos, P. and Karakostas, V. (1995). *System Requirements Engineering*. McGraw-Hill International.

Low, J. and Woolgar, S. (1993). 'Managing the Social–Technical divide: some aspects of the discursive structure of information system development'. *CRICT Discussion Paper 33, Brunel University*.

Lubars, M., Potts, C. and Richter, C. (1993). 'A Review of the State of the Practice in Requirements Modeling'. *Proceedings of IEEE International Symposium on Requirements Engineering, San Diego, California*, pp. 2–15. Los Alamitos, Cal: IEEE Computer Society Press.

Luff, P., Heath, C. and Greatbach, D. (1994). 'Work, interaction and technology: The naturalistic analysis of human conduct and requirements analysis', in M. Jirotka and J. Goguen (eds), *Requirements Engineering: social and technical issues*, pp. 259–289. San Diego, Cal.: Academic Press.

Luff, P., Jirotka, M., Heath, C. and Greatbach, D. (1993). 'Tasks and social interaction: the relevance of naturalistic analyses of conduct for Requirements Engineering'.

Proceedings of IEEE International Symposium on Requirements Engineering, San Diego, California, pp. 187–92. Los Alamitos, Cal: IEEE Computer Society Press.

Lutz, R. R. (1993). 'Analysing software requirements errors in safety-critical, embedded systems'. *Proceedings of IEEE International Symposium on Requirements Engineering, San Diego, California*, pp. 126–34. Los Alamitos, Cal: IEEE Computer Society Press.

Lyytinen, K. and Hirschheim, R. (1987). *Information Systems Failures – a survey and classification of the empirical literature.* Oxford Surveys in Information Technology, Vol 4, pp. 257 – 309. Oxford University Press.

Macaulay, L. A. (1993). 'Requirements as a Cooperative Activity'. *Proceedings of IEEE International Symposium on Requirements Engineering, San Diego, California.* Los Alamitos, Cal: IEEE Computer Society Press.

Macaulay, L. A. (1995a). *Report on a survey of requirements for Requirements Engineering courses.* (Internal report.) Department of Computation, UMIST, PO Box 88, Manchester, M60 1QD, UK.

Macaulay, L. A. (1995b). *Human Computer Interaction for Software Designers.* Thomson International Press.

Macaulay, L. A., (1994). 'Cooperative Requirements Capture: Control Room 2000', in M. Jirotka and J. Goguen (eds), *Requirements Engineering: Social and Technical Issues*, pp. 67–87. Academic Press.

Macaulay, L. A., (1995c). 'Cooperation in Understanding User Needs and Requirements'. *International Journal of Computer Integrated Manufacturing Systems* 8(2): 155–65.

Marca, D. and McGowan, C. (1988). *SADT: Structured Analysis and Design Technique.* New York: McGraw-Hill.

Markus, M. and Bjorn-Anderson, N. (1987). 'Power over Users: Its Exercise by System Professionals'. *Communications of the ACM* 30.

Marsh, S. (1991). 'Facilitating and training in Quality Function Deployment'. Methuen, Mass.: GOAL/QPC.

McMaster, T., Jones, M. C. and Wood-Harper, A. T. (1995). 'Implementation Planning: A Role for an "Information Strategy"'. *Third Decennial Conference Computers in Context: Joining Forces in Design, Aarhus, Denmark.* Matematisk Institut, Aarhus Universitet, Trykkeriet.

McMenamin, S. and Palmer, J. (1984). *Essential Systems Analysis.* Englewood Cliffs, NJ: Prentice-Hall.

Mills-Packo, P. A., Wilson, K. and Rotar, P. (1991) 'Highlights from the use of the soft systems methodology to improve agrotechnology transfer in Kona, Hawaii'. *Agricultural Systems* 36: 409–25.

Mitroff, I. I. (1980). *Management myth information systems revisited: a strategic approach to asking nasty questions about system design.* Amsterdam: The Human Side of Enterprise.

Monk, A. F. and Curry, M. B. (1994). 'Discount dialogue modelling with Action Simulator', in G. Cockton, S. W. Draper and G. R. S Weir (eds), *People and Computers 9 – proceedings of HCI'94.* Cambridge University Press.

Monk, A., Wright, P., Haber, J. and Davenport, L. (1993). *Improving Your Human Computer Interface: A practical technique.* Prentice-Hall International.

Morgan, D. L. (1988). *Focus groups as qualitative research.* Sage.

Moser, C. and Kalton, G. (1971). *Survey Methods in Social Investigation.* Gower.

Mullery, G. P. (1987). 'CORE – a method for controlled requirements expression', in R. H. Thayer and M. Dorfman (eds), *System and Software Requirements Engineering*, pp. 304–13. Los Alamitos, Cal: IEEE Computer Society Press.

Mumford, E. (1986). *Designing Systems for Business Success, the ETHICS Method.* Manchester: Manchester Business School.

Mumford, E. (1989). 'User Participation in a Changing Environment – Why we need it', in K. Knight (ed.), *Participation in Systems Development*, pp. 60–72. Kogan Page.

Neale, I. M. (1989). 'First generation expert systems: a review of knowledge acquisition methodologies'. *The Knowledge Engineering Review.*

Newman, W. M. and Lamming, M. G. (1995). *Interactive System Design.* Berkshire, UK: Addison-Wesley.

NUPE (1993). 'An accident waiting to happen'. NUPE's submission to the *London Ambulance Enquiry.*

Ohnishi, A. and Agusus, K. (1993). 'CARD: A Software Requirements Definition Environment'. *Proceedings of IEEE International Symposium on Requirements Engineering, San Diego, California.* Los Alamitos, Cal: IEEE Computer Society Press.

Orr, K. (1980). 'Structured Requirements Definition in the 80s'. *ACM'80 Conference,* pp. 350–4. New York: ACM Press.

Pape, T. and Thoreson, K. (1987). 'Development of common systems by prototyping', in G. Bjerknes, P. Ehn and M. Kyng (eds), *Computers and Democracy – a Scandinavian Challenge,* pp. 297–311. Aldershot, UK: Avebury.

Pasch, J. (1991). 'Dialogical software design', in H.-J. Bullinger (ed.), *Human Aspects in Computing: Design and Use of Interactive Systems and Work with Terminals,* pp. 556–60. Amsterdam: Elsevier.

Pohl, K. (1993). 'The three dimensions of Requirements Engineering', in C. Rolland, F. Bodart and C. Cauvet (eds), *Fifth International Conference on Advanced Information Systems Engineering (CAiSE'93),* pp.175–292. Paris: Springer-Verlag.

Preece, J., Rogers, Y., Sharp, H., Benyon, D., Holland, S. and Carey, T. (1994). *Human–Computer Interaction.* Addison-Wesley.

Reich Y., Konda S., Monarch I. and Subrahmanian E. (1992). 'Participation and Design: An Extended View', in *Proceedings of the Participatory Design Conference PDC'92, Boston, MA.*

Robinson, B. (1994). Social Context and Conflicting Interests. *Second BCS Conference on Information Systems Methodologies, Edinburgh,* pp. 235–49. British Computer Society.

Ross, D. T. (1977). 'Structured Analysis: A Language for Communicating Ideas'. *IEEE Transactions on Software Engineering* 3(1): 16–34.

RTM (1994). Marconi Systems Technology. Requirements and Traceability Management Tools.

Rubin, J. (1994). *Handbook of Usability Testing: How to plan, design, and conduct effective tests.* John Wiley and Sons.

Rzevski, G. (1980). 'On the design of a design methodology', in R. Jacques and J. A. Powell (eds), *Design: Science: Method.* Westbury House.

Schuler, D. and Namioka, A. (eds) (1993). *Participatory Design: Principles and Practices.* Hillsdale, NJ: Lawrence Erlbaum.

Skousen, T. (1982). 'System Development in a shared data environment, the D2S2 methodology', in T. W. Olle *et al.* (eds), *Information system design methodologies: a comparative review.* North-Holland.

Smith, T. J. (1993). READS: 'A Requirements Engineering Tool'. *Proceedings of IEEE International Symposium on Requirements Engineering, San Diego, California,* pp. 94–8. Los Alamitos, Cal: IEEE Computer Society Press.

Sommerville, I., Rodden, T., Sawyer, P., Bentley, R. and Twidale, M. (1993). 'Integrating Ethnography into the Requirements Engineering Process'. *Proceedings of IEEE International Symposium on Requirements Engineering, San Diego, California,* pp. 165–74. Los Alamitos, Cal: IEEE Computer Society Press.

Souquieres, J. and Levy, N. (1993). 'Description of Specification Developments'. *Proceedings of IEEE International Symposium on Requirements Engineering, San Diego, California,* pp. 216–25. Los Alamitos, Cal: IEEE Computer Society Press.

Sowa, J. F. (1984). *Conceptual Structures: Information Processing in Mind and Machine.* Addison-Wesley.

Strain, J. D. (1990). 'A harder approach to soft systems: matching men and technology in warship systems'. *Journal of Applied Systems Analysis* 17: 123–32.

Suchman, L. and Jordan, B. (1990). 'Interactional troubles in face-to-face survey interviews'. *Journal of the American Statistical Association* 85(409): 232–41.

Sullivan, L. P. (1986). 'Quality function deployment'. *Quality Progress* 19(6): 39–50.

Taylor, B. (1990). 'The HUFIT planning analysis and specification toolset', in D. Diaper, D. Gilmore, G. Cockton and B. Shackel (eds), *INTERACT'90*, pp. 371–6. Cambridge: IFIP.

Thompson, D. M. M. and Fallah, M. H. (1989). 'QFD – A Starting Point for Customer Satisfaction Metrics'. *IEEE International Conference on Communications, Boston, June 1989, IEEE.*

Tudor, D. J. and Tudor, I. J (1995). 'Systems Analysis and Design: A comparison of structured methods', *NCC.* Oxford: Blackwell.

Van Lieshout, M. and Massink, M. (1993). 'Constructing a vulnerable society', in Berleur, Beardon and Laufer (eds), *Facing the challenge of risk and vulnerability in an information society.* North-Holland.

Van Schouwen, A. J., Parnas, D. L. and Madey, J. (1993). 'Documentation of Requirements for Computer Systems'. *Proceedings of IEEE International Symposium on Requirements Engineering, San Diego, California*, pp. 198–208. Los Alamitos, Cal: IEEE Computer Society Press.

Verheijen, G. M. and Van Bekkum, J. (1982). 'NIAM: An information analysis method', in T. W. Olle *et al.* (eds), *Information system design methodologies: a comparative review.* North-Holland.

Vidgen, R. (1994). 'Research in progress: using stakeholder analysis to test primary task conceptual models in information system development', in Lissoni, C. *et al.* (eds), *Proceedings of the second conference on Information Systems Methodologies*, pp. 223–34. BCS ISM Specialist Group.

Viller, S. A. (1990): *Computer Support for Group Facilitators: An Investigation.* (MSc. Thesis.) University of Manchester.

Viller, S. A. (1991a). 'The Group Facilitator: a CSCW Perspective', in L. Bannon, M. Robinson and K. Schmidt (eds), *ECSCW'91. The Second European Conference on Computer Support for Cooperative Work, Amsterdam, Holland.* Kluwer.

Viller, S. A. (1991b). *CRC/211-Group Support.* (Internal project report from the Cooperative Requirements Capture project.) Department of Computation, UMIST, PO Box 88, Manchester, M60 1QD, UK.

Wallmuller, E. (1994). *Software Quality Assurance: A Practical Approach.* Prentice-Hall International.

Ward, P. and Mellor, S. (1985). *Structured Development for Real-Time Systems.* Englewood Cliffs, NJ: Prentice-Hall.

Wasserman, A. I., Freeman, P. and Porcella, M. (1983). 'Characteristics of software design methodologies', in T. W. Olle *et al.* (eds), *Information system design methodologies: a feature anaysis.* North-Holland.

Watson, R. and Smith, R. (1988). 'Applications of the Lancaster soft systems methodology in Australia'. *Journal of Applied Systems Analysis* 15: 3–26.

Westley, F. and Walters, J. A. (1988). 'Group facilitation skills for managers'. *Management Education and Development* 19(2): 134–43.

Wilson, B. (1984). *Systems: Concepts, methodologies and applications.* (2nd edition 1990.) Chichester: John Wiley and Sons.

Woodson, W. E. (1981). *Human Factors Design Handbook: Information and Guidelines for the Design of Systems, Facilities, Equipment, and Products for Human Use.* New York: McGraw-Hill.

Yourdon, E. (1989). *Modern Structured Analysis.* Englewood Cliffs, NJ: Yourdon Press.

Zultner, R. E. (1989). 'Software Quality [Function] Deployment: Applying QFD to software'. *The second symposium on Quality Function Deployment*, GOAL/QPC (op. cit.).

Zultner, R. E. (1993). 'TQM for Technical Teams'. *Communications of the ACM* 36(10): 79–91.

SUBJECT INDEX

AUTHOR INDEX